ALUM

More Precious than Gold

For James, Adnan and Zoe

ALUM
MORE PRECIOUS THAN GOLD

The Amazing Influence and Trade of Alum Across the Ages

SYDNEY THORNE

PEN & SWORD HISTORY
AN IMPRINT OF PEN & SWORD BOOKS LTD
YORKSHIRE – PHILADELPHIA

First published in Great Britain in 2024 by
PEN AND SWORD HISTORY
An imprint of
Pen & Sword Books Ltd
Yorkshire – Philadelphia

ISBN 978 1 03610 108 4

A CIP catalogue record for this book is available from the British Library.

Typeset in Times New Roman 11.5/15 by
SJmagic DESIGN SERVICES, India.
Printed and bound in the UK by CPI Group (UK) Ltd.

Pen & Sword Books Limited incorporates the imprints of After the Battle, Atlas, Archaeology, Aviation, Discovery, Family History, Fiction, History, Maritime, Military, Military Classics, Politics, Select, Transport, True Crime, Air World, Frontline Publishing, Leo Cooper, Remember When, Seaforth Publishing, The Praetorian Press, Wharncliffe Local History, Wharncliffe Transport, Wharncliffe True Crime and White Owl.

For a complete list of Pen & Sword titles please contact
PEN & SWORD BOOKS LIMITED
George House, Units 12 & 13, Beevor Street, Off Pontefract Road,
Barnsley, South Yorkshire, S71 1HN, England
E-mail: enquiries@pen-and-sword.co.uk
Website: www.pen-and-sword.co.uk

or

PEN AND SWORD BOOKS
1950 Lawrence Rd, Havertown, PA 19083, USA
E-mail: uspen-and-sword@casematepublishers.com
Website: www.penandswordbooks.com

FSC
www.fsc.org
MIX
Paper | Supporting responsible forestry
FSC® C013604

Contents

Acknowledgements

Thank you to Andrew Thorne and Nicky Thorne for reading and giving sound advice on improving the manuscript.

And thank you, Irmgard and all the family, for all your support and for patiently listening to my endless tales of alum.

Maps

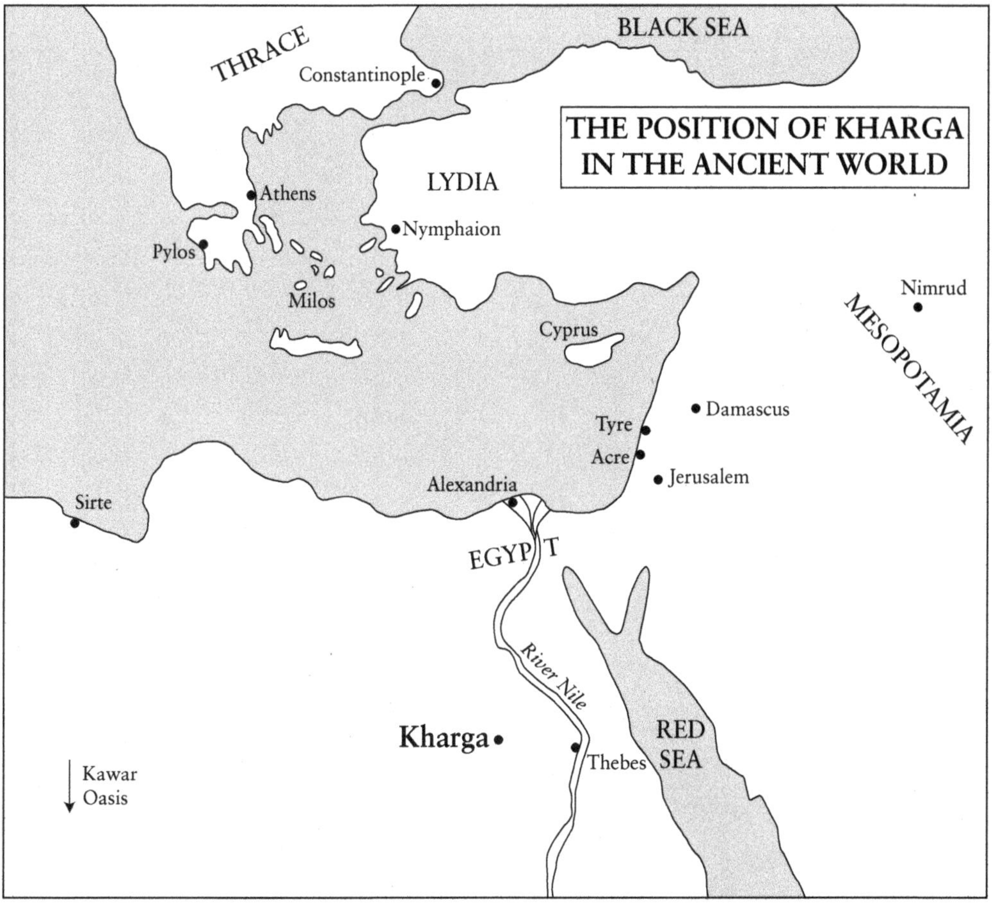
THE POSITION OF KHARGA IN THE ANCIENT WORLD
BLACK SEA
THRACE
Constantinople
LYDIA
Athens
Nymphaion
Pylos
Milos
Nimrud
MESOPOTAMIA
Cyprus
Damascus
Tyre
Acre
Jerusalem
Alexandria
Sirte
EGYPT
River Nile
Kharga
Thebes
RED SEA
Kawar Oasis

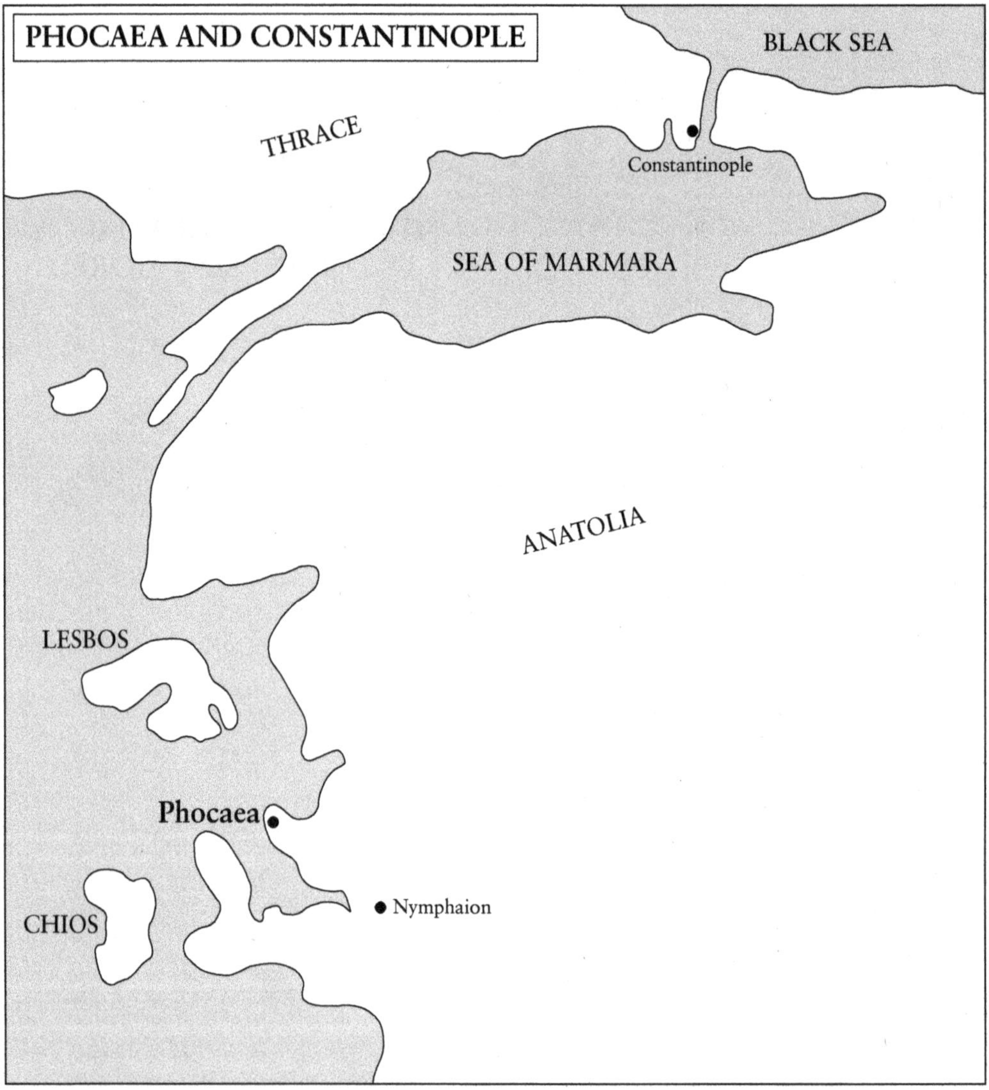
PHOCAEA AND CONSTANTINOPLE
BLACK SEA
THRACE
Constantinople
SEA OF MARMARA
ANATOLIA
LESBOS
Phocaea
Nymphaion
CHIOS

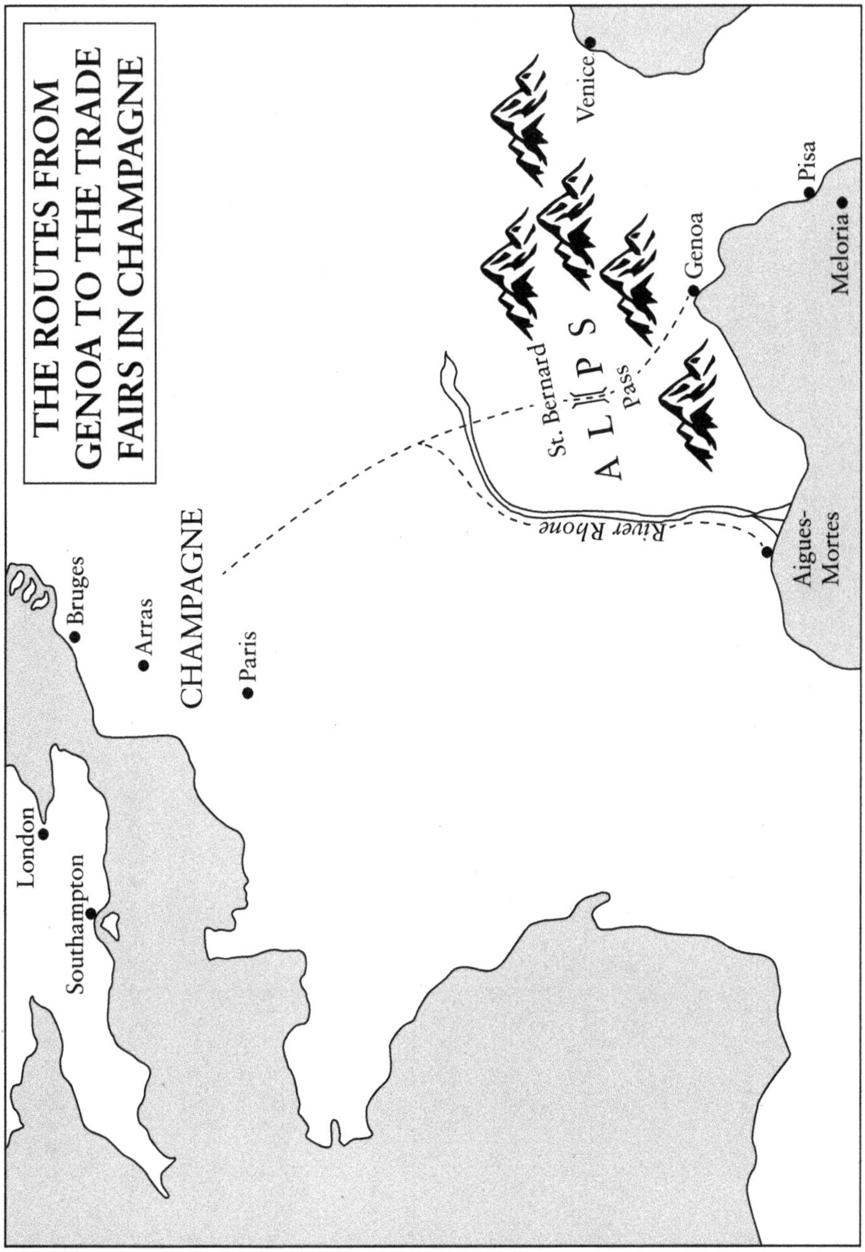
THE ROUTES FROM
GENOA TO THE TRADE
FAIRS IN CHAMPAGNE
London
Southampton
Bruges
Arras
CHAMPAGNE
Paris
CHAMPAGNE
St. Bernard
Pass
ALPS
River Rhone
Aigues-
Mortes
Genoa
Venice
Pisa
Meloria

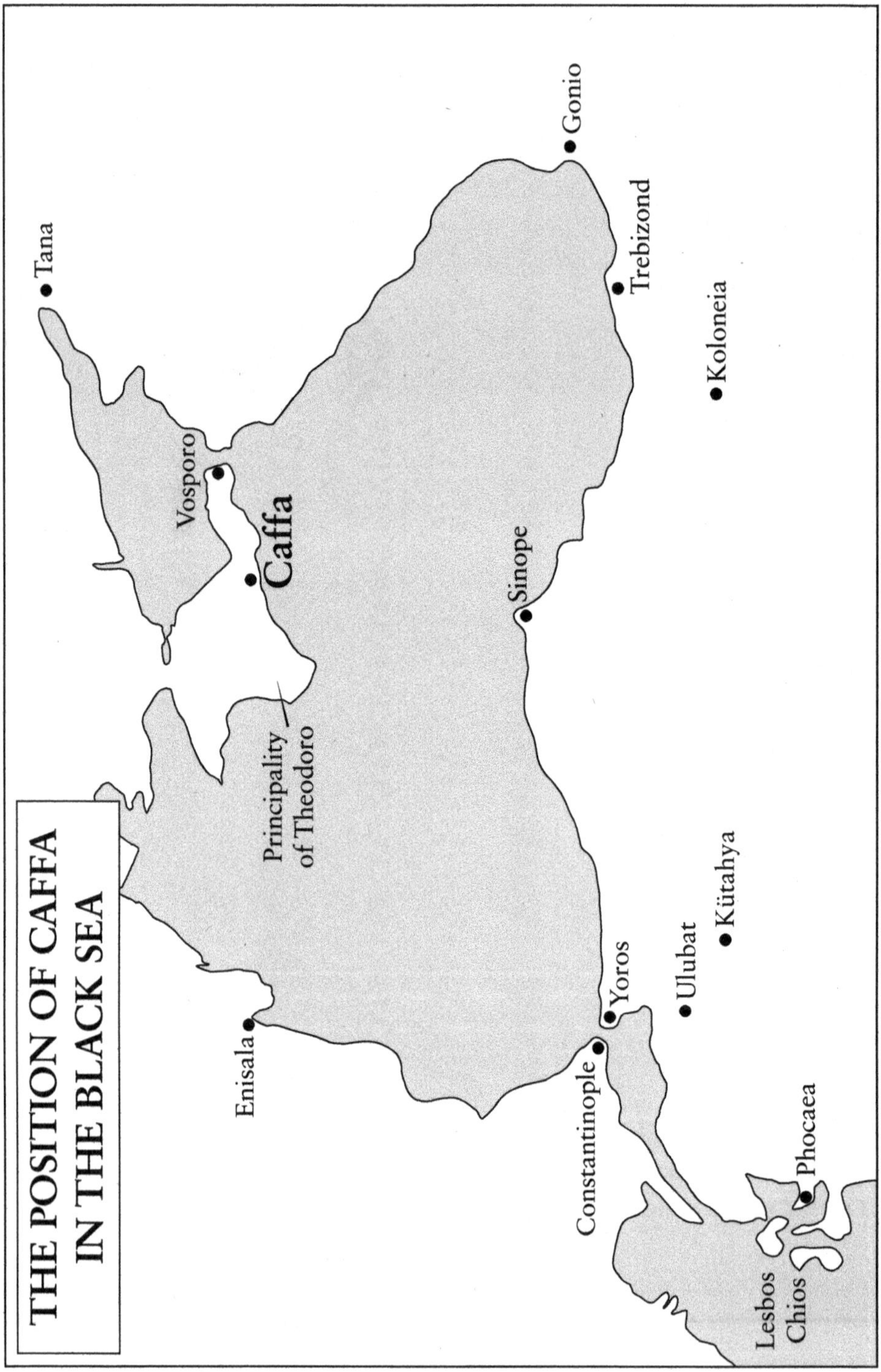
THE POSITION OF CAFFA
IN THE BLACK SEA
Tana
Vosporo
Caffa
Principality
of Theodoro
Enisala
Sinope
Gonio
Trebizond
Koloneia
Kütahya
Ulubat
Yoros
Constantinople
Phocaea
Lesbos
Chios

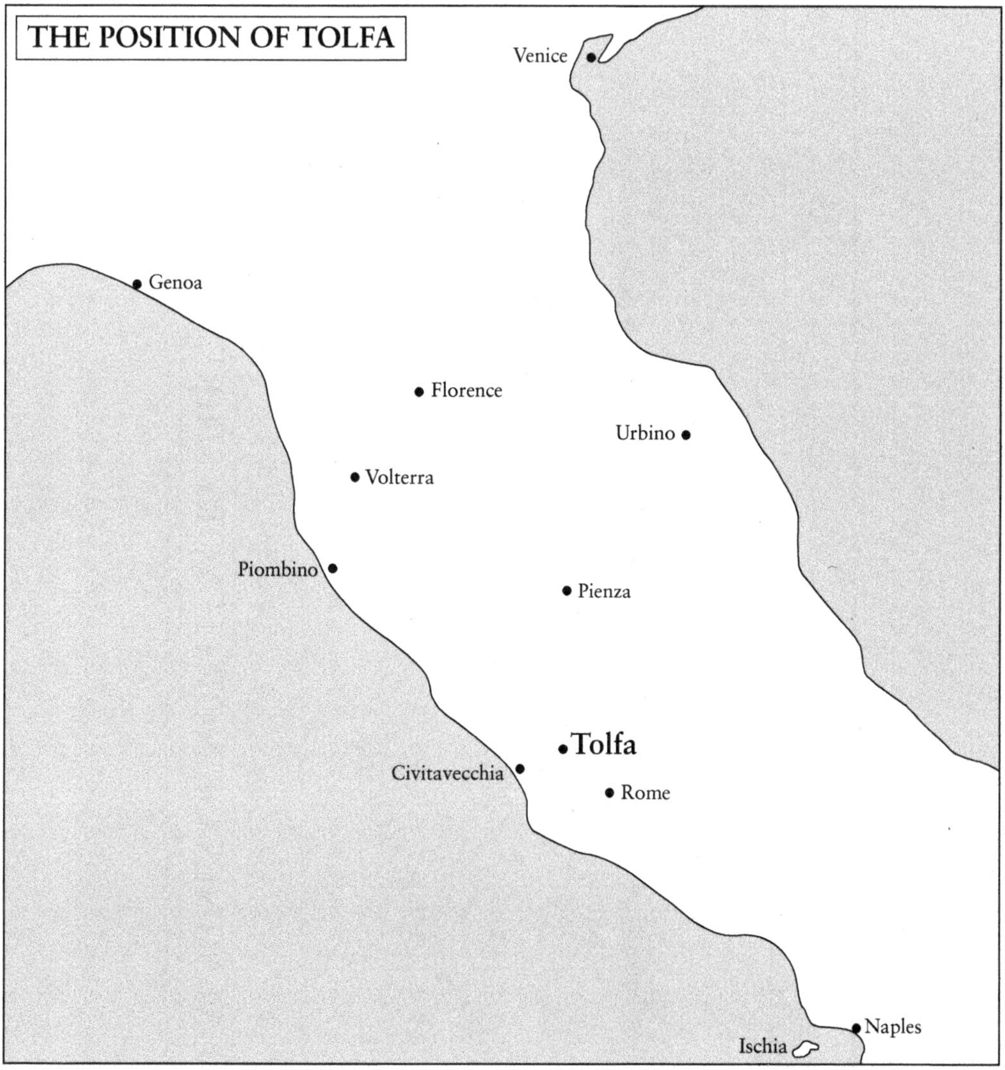
THE POSITION OF TOLFA
Venice
Genoa
Florence
Urbino
Volterra
Piombino
Pienza
Tolfa
Civitavecchia
Rome
Naples
Ischia

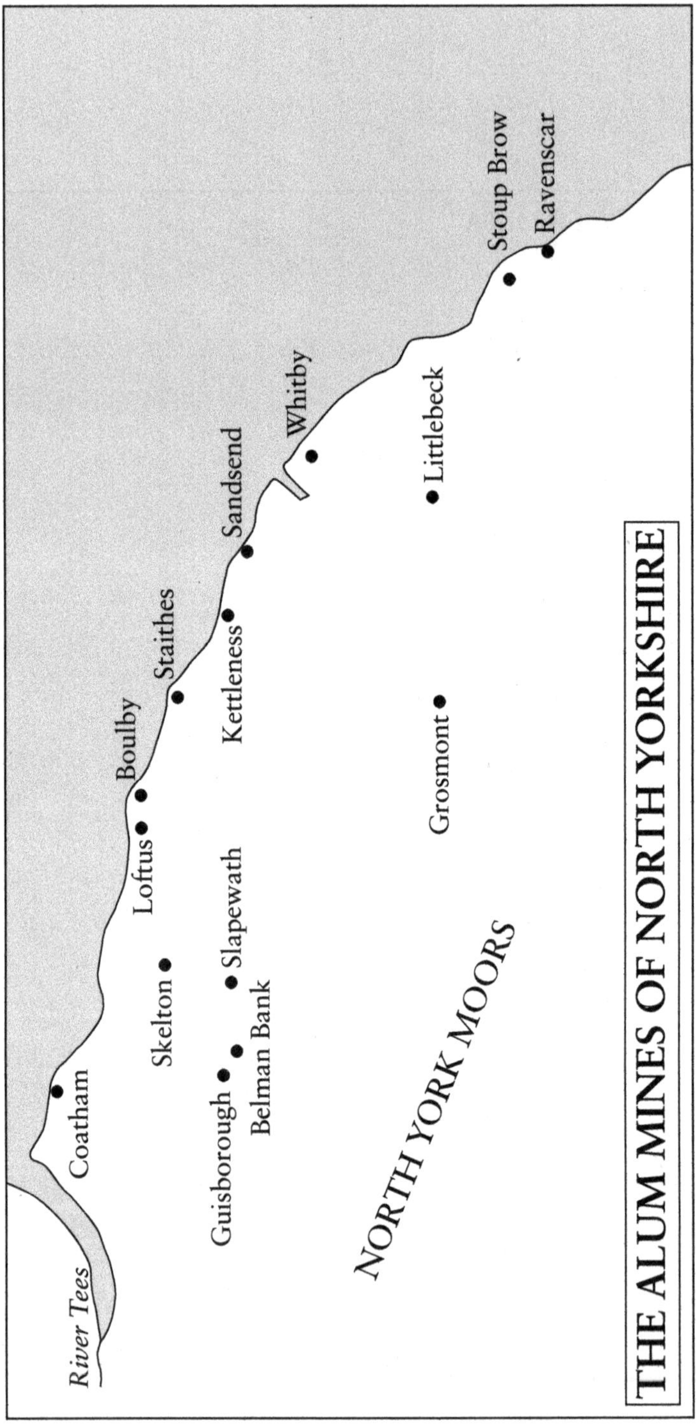

THE ALUM MINES OF NORTH YORKSHIRE

1

Alum? Never Heard of It!

It was everywhere, yet always curiously out of sight.

The Egyptians traded it from at least 1500 BCE. It made the Eastern Roman Empire so rich that it could pay off its enemies and survive for 1500 years. It gave Saladin and his successors the means to see off the crusaders. The Chinese emperors claimed a monopoly on it. The Ottomans used the proceeds from its sales to fund their fabled palaces. The Genoese built a Mediterranean empire on it. The pope thanked God for it when it was discovered near Rome in 1461. It would, he was assured, give him 'victory over the Turk'.

In England, it was extracted so frenziedly that millions of tons of rock were ripped from the cliffs on the Yorkshire coast. Factories were put up, armies of workers were employed, harbours were built and ingredients chemically treated on an industrial scale – fully 150 years before the Industrial Revolution. It was, said the French chemist Jean Antoine Chaptal as late as 1787, more valuable than gold or silver.[1]

What was this highly desirable 'it'?

It was alum.

No, not aluminium. Alum.

OK, so you have never heard of the stuff. You're not alone. Most people haven't. Yet tanners used alum to tan leather. Potters used alum to colour their pots. Ironsmiths used alum to temper the iron blades of knives and swords, and physicians to stop bleeding. Goldsmiths used alum to refine real gold and crooks used alum to forge fake gold. Alum was used to make paper on which ink would not smudge, and to make invisible ink. The Venetians used it in the manufacture of their famous glass. Alum was used to refine sugar, preserve food, and remove stains.[2] On wood, it was a fire-retardant. It cleared cloudy water, making it pleasant to drink. Mixed with mastic, it was used to fill dental cavities.[3]

And that was not even the half of it. The *real* reason why alum was for centuries an essential commodity in the world economy was its vital role in what was for centuries the world's most important industry: the production of woollen cloth.[4]

Mass-produced factory-made cloth is so cheap and dispensable today that it is almost impossible for us to think of woollen cloth as a highly valuable product. But for hundreds of years, that is what it was, and as a result, the manufacture of woollen cloth underpinned the wealth of urban centres across the world, from Flanders and Italy to the countries of the Middle East and to Persia, India and China.

Everyone, of course, needed cloth. Yet as every stitch of every piece of cloth was woven by hand, even small pieces took hours to make and were therefore, compared to the price of cloth today, phenomenally expensive. As for larger pieces of cloth such as bed sheets, they were often more valuable than the wooden chests they were stored in, and were individually assigned in wills as treasured assets that one generation passed on to the next.

Good quality fleeces, shorn and carded by hand, fetched such a high price that the Lord Chancellor's seat in Parliament in London was (and indeed still is) not a chair but a cushion packed with wool: anyone with so much wool at their command, it was understood, had no need for bribes. Spin the wool and weave it into cloth, and its value increased. The wealth of the cloth trade is reflected in the magnificent, richly decorated medieval cloth halls of Bruges, Ghent, Ypres and other cities in Flanders, as opulent and impressive as any church or cathedral.

Now, add colour to the cloth, and its value rose again. That's why only the rich could afford coloured cloth, and they flaunted colour shamelessly, in ladies' shawls and dresses, in men's tunics, doublets and cloaks, in cushions, curtains, tablecloths and other soft furnishings, in pennants and draperies, and even in the sails of their merchant ships. As much as gold or jewellery, coloured cloth was public proof of status.

But why was colour so expensive? There was no lack of natural dyes. Red came from the roots of madder, which grew abundantly in southern Europe, North Africa and Central Asia.[5] Blue was obtained from the leaves of woad, which grew over an even larger area. Onion skin, dyers' rocket and dyers' chamomile produced yellow, lichen produced violet, and violet lilac flowers boiled with foxglove gave green. So did woad used with a yellow dye. The possibilities for dyers were endless.

But here's the rub. Most of these colours weren't 'fast'. The dyes ran if the cloth came into contact with water. And they faded in the sun. So if the colours were not to leach out in the wash, or simply the first time the cloth was worn in the rain or sun, dyers had to find a product that would fix the colour indelibly to the cloth. And the best fixative, or *mordant*, better than any mixtures made from bark and leaves, was alum.[6]

Alum opens the scales of the wool, giving the dye better access to its fibres.[7] Use alum in the dyeing process, and your colours were safe. You could wear your clothes and flaunt your wealth with confidence. The colours would not run. No wonder alum commanded a high price.

And alum did more than only fix the dye. Alum gave the colours a deeper, more luxuriant shine. So the more the rich cried out for colour, the louder the dyers and cloth makers cried out for alum. It was, said Italian scientist Vanoccio Biringucci in the sixteenth century, as 'essential to dyers as bread is to man.'[8]

For at all times and in all cultures the rich and the powerful in government, in religion, in battle, have always sported the most spectacularly colourful clothes. The Mughal and the Aztec courts radiated sumptuous colour. Europe's Renaissance and Tudor courts flaunted ostentatious colour. The Chinese and Roman emperors each reserved a colour for themselves – yellow for the former, purple for the latter.[9] Elites across the world were so obsessed with colour that they passed *sumptuary laws* which limited the amount of colour that the common people were allowed to use. The rich could then keep the brightest colours for themselves. And alum was the magic ingredient that made these colours possible.[10]

Even that outward sign of humility, the pure white in the cassocks and headwear of priests, imams and elders in so many of the world's religions, could only be achieved with alum.

No wonder, then, that there was a high demand for alum. But there were few sources of it. For centuries, you could only get alum in and around the Mediterranean. Far from English sheep, which produced the best wool.[11] Far from Flanders weavers, who wove the best cloth. So alum had to be bought, and shipped, and traded. Corner the market and you could become immensely rich.

What, then, is alum exactly? Why have most people never heard of it? And why, if it underpinned the wealth of societies across the world, does alum get so few mentions in history books?

Alum is a salt made of two different sulphate compounds: aluminium sulphate and the sulphate of either potassium or ammonia.[12] In its pure state alum is a crystal or a powder, which, crucially for the dyeing process, is soluble in water. Small quantities of alum crystals occur naturally, but most alum was obtained in an elaborate process of burning and soaking a rock called *alunite* or, later, a shale called *alum shale*, or *schist*.

Those who extracted the alum and traded in it could make a fortune. But unlike silks and spices, pearls, gold and exotic wood, alum was not a commodity that found pride of place in rich people's houses. The people who benefitted from alum in their sumptuous garments and tapestries had no particular interest in how the colour in their clothes was created and secured. They never saw or handled alum. And late medieval and Renaissance painters who delighted in every pleat of luxuriantly coloured cloth made possible by alum, never painted alum as such.

So alum remained the hidden ingredient, known only to those who produced and traded it. It was the secret of dyers, tanners and other craftsmen who used it, and of the merchants who bought and sold it. Proof of its use was there in every yard of coloured cloth, in the leather used to make belts and bind books, but the alum itself remained unseen. It is invisible to archaeologists too. Gold, iron, pottery or even wood can be excavated, but alum, being a salt, decays. It literally disappears from the face of the earth.

And as for not featuring in history books, even an alum nerd such as myself has to admit that set against commodities such as pepper, pomegranates and pineapples, saffron and cinnamon, silk, satin and damask, alum is spectacularly unsexy. When traded it was often carried as ballast in the smelly bilges of ships. One of the ingredients used to extract alum from alunite or shale was human urine: the urine was brought in vast quantities by cart or boat in evil-smelling tubs, and the smell did not improve in storage. And let's face it, alum *schist* is not a pretty word. These are not the attributes of a commodity that artists were eager to paint or historians to investigate.

As to why there is so little trade in alum today, the reason is simple. Cheaper synthetic mordants invented in the nineteenth century, along with synthetic dyes which contained their own fixatives spelled the end of the alum manufacturing industry and trade. Alum is still an ingredient in styptic pencils to staunch bleeding from minor cuts, and is peddled as an astringent aftershave or skin whitener. But it no longer plays the pivotal role in the rich economies of the world that it once did. As the abandoned alum works and quarries have decayed, therefore, alum has disappeared from public consciousness. We admire the flamboyant coloured fabrics in Chinese and Persian prints and the luminous cloaks and dresses in Flemish and Italian paintings, but alum, the ingredient that made these colours possible, has slipped from memory.

Still, alum had a good run for its money. It was extracted, traded and prized for three thousand years and more. And this book will tell some of its fascinating stories.

2

Kharga: Alum in the Desert

Look at a map of the ancient world, and the Egyptian oasis of Kharga doesn't exactly leap at you as a promising location for international trade.

Kharga lies deep in Egypt's *western* desert, the wrong side of the Nile for easy access to the flourishing markets of Persia and Mesopotamia. Thousands of miles of relentlessly arid country lie to the north, south and west, while to the east some 120 miles of desert and cliffs, a thousand feet high, separate it from the irrigated fields of the Nile valley. A glance at the weather chart shows that temperatures typically hit up to 50^0 C in June and fall to 0^0 C in January. An unlikely setting, then, for what has been called a 'rich, cultural crossroad'.[1]

But ancient Kharga possessed unexpected assets. Its size, for one, comes as a surprise. I had never imagined that a desert oasis could be 100 miles long and up to 22 miles wide, as Kharga is today; and back in 1000 BCE, it was very much larger than that, with ample space for growing food to feed a large population.[2] There were vineyards too, with Kharga's wine so highly prized that a jar of it was found inside the tomb of pharaoh Tutankhamun himself.[3] A large lake at the heart of the oasis supplied water for a community that included not only farmers, peasants and cattle breeders[4], but craftsmen, soldiers, merchants and priests. The latter served in temples such as the monumental Temple of Hibis, built in 500 BC on the site of an even earlier temple.[5] Kharga was indeed a site of ancient culture.

What's more, far from being isolated, ancient Kharga was in fact well connected[6] across the desert, thanks to a revolution in transport more fundamental than the narrow-gauge Western Oasis railway line built by the British in 1907, or the tarmac roads that now link Kharga to the Nile valley: the domestication of the dromedary and camel. From at least the 9th century BCE camels replaced the far less suitable donkeys as draught animals in the desert, facilitating cross-desert trade on a scale that had never been possible before. Camels needed fewer water holes than donkeys and could be loaded with more goods, as they didn't also have to carry their own food and water.[7] Soon caravans of camels were carrying gold, ivory and ostrich feathers on the Darb el-Arba'in trail from Nubia and Darfur across the Sahara to Kharga, and from there to the ports in the Nile Valley or on the Red Sea. On this last stretch of the trail, the camels carried Kharga's local goodies too, of course: its award-winning wine, its salt (used in preserving food), its natron (used to mummify corpses) – and its alum.[8]

Natural crystals[9] of alum were easy enough to find all around the lake in the centre of the Kharga oasis. The lake was retreating in the desert heat just as the Dead Sea is shrinking today, and as the waters evaporated they left behind mineral crystals, many of which were alum crystals. It was easy enough to scrape the crystals off the ground – but who, I wonder, first had the astonishing idea of using them to fix colour in cloth? And did this discovery originate in Egypt or in some other country where alum crystals were available?[10] We do not know. Some think that tanners, who made raw pelts into leather, were the first in on the act. Tanners had found that if they tanned their hide with alum (the process was then called 'tawing') it made the leather softer, more supple and easier to dye or stain than if they tanned it with vegetables, or cured it with oil or smoke. It may be that dyers copied their idea.

What we do know from the archaeological evidence is that dyers in Egypt were using alum as a mordant with madder before 1500 BCE,[11] maybe even as far back as 2000 BCE,[12] to dye cloth red. Alum was also being used in medicine: medical papyri from about 1500 BCE mention alum twenty-six times, mostly in the context of staunching bleeding from wounds. For a pulled rib, the recommended treatment was to apply a bandage lined with alum before treating the wound daily with honey.[13]

Craftsmen also made use of alum in the decorative arts. Statuettes laid in a boiling solution of alum, copper alloys and vinegar acquired a deep black sheen, or *patina*, which could then be beautifully set off by inlaying the surfaces with tiny pieces of gold or silver. A crocodile in the Egyptian Museum in Munich is a good example: look at it and it is hard to believe that such a sophisticated technique was being used four thousand years ago.[14] That was the power of alum.

Then, when the New Kingdom[15] (the one with such famous names as Tutankhamun, Nefertiti and Ramesses II) brought in a new taste for glass and ceramics with a distinctive cobalt blue, this colour too, was achieved by using alum. It was likely from the Kharga oasis, for alum salts from Kharga were rich in cobalt.[16] Inscriptions at Ramesses' great palace at Luxor, built in the 13th century BCE, list Egypt's most important resources and their origin, and sure enough Kharga and its alum are included.[17]

Of course, a mineral with so many uses in dyeing, tanning, medicine and the decorative arts was soon in demand across the ancient world, putting Kharga firmly on the map of international trade. Inscriptions on tablets from 1600-1200 BCE in the city of Pylos in southern Greece record imports of alum: one load was exchanged for today's equivalent of 130 kilos of wool and 10 cloths, and another for six kilos of wool, four sheep, three cloths, 288 litres of wine and 384 litres of figs.[18] A tablet from about 1000 BCE, inscribed in cuneiform script, records the purchase in Mesopotamia of 233 minas of alum, for which a price of 1 mina 17 2/3 shekels was paid.[19] And in this case, the country of origin was specified: the alum came from Egypt.[20]

In 879 BCE, when King Ashurnasirpal II of Assyria summoned craftsmen from all over his empire to furnish the palaces in his new capital at Nimrud,[21]south of present-day Mosul, he made sure there was at least one alum worker among them so he could be sure, when he threw a party for nearly 70,000 guests to inaugurate his new city, that the colours in the regal garments and in the palaces' soft furnishings were both bright and fast. An Assyrian text even gives us precise measures of how much alum to use in the dyeing process: cushions, we are told, are dyed with 4 2/3 minas of alum to 14 minas of madder.[22]

Then, when the Assyrians invaded Egypt in 671 BCE, they seized control of the alum-rich oasis of Kharga, as well as the Nile Valley. The historian Herodotus gives an example of the long-distance trade that existed in this period: when the temple of Apollo in Delphi was accidentally destroyed in a fire in 548 BCE, pharaoh Amasis II sent 26 tons (!) of Egyptian alum to help in its reconstruction.[23] Trade in Egyptian alum in the sixth century BCE is confirmed in documents unearthed in Mesopotamia.[24]

Six generations later the Persians were on the move. First, in 546 BCE, they defeated King Croesus of Lydia, a country (in what is now the west of Turkey) that has faded from memory, while the wealth of its king was so stupendous that the saying 'as rich as Croesus' has survived. Which is pertinent to our story, for Croesus's legendary wealth was down to yet another wonderful use of alum: when the Lydians heated an alloy of gold, copper and silver, which they found in their river beds, with salt and alum to nearly 800^{0}C, the copper and silver separated out – leaving the Lydians with pure gold.[25] Thanks to alum, therefore, you could trust Lydian gold to be pure, and King Croesus of Lydia became the world's first monarch to issue gold coinage.

Having taken Lydia, the Persians conquered Babylon and, in 525 BCE, Egypt.[26] It was the Persians then, who built Kharga's Temple of Hibis in 500 BCE, and the Persian king Darius I is depicted on its walls. Crucially for our story of alum, however, the Persians brought their tunnelling skills, and they were astonishing.

Over the centuries the Persians had developed the engineering know-how to dig tunnels that tapped sources of underground water, and were angled so that the water flowed down to where it was needed in fields and gardens. These tunnel systems, called *qanats*, could be on a massive scale. One of the longest, dug in Zarch in central Persia in 1000 BCE, stretched an extraordinary seventy miles!

By the time the Persians arrived in Kharga, the lake at the centre of the oasis had retreated further, so the conquerors tunnelled through the rock to improve the water supply.[27] Here too they dug on a grand scale, for when one of their *qanats* in Kharga was cleared of debris in the early 1900s, it was found that the Persians had hand-shifted some eleven thousand tons of rock to create it. What's more, the *qanat* was so well built that, once it was cleared, water flowed through it as it had done over two thousand years before.[28]

With such skills at their disposal, it is likely that the Persians dug Kharga's first alum mines, or vastly extended any rudimentary mines that already existed. Mining had become necessary because the high demand for alum had already exhausted the naturally occurring alum crystals around the lake. When the British geologist and Egyptologist Llewellyn Beadnell explored Kharga in the early 1900s, he found that its hills were 'literally honeycombed with ancient workings', many of which 'penetrate for long distances underground'. He found 'huge dump heaps' at the entrance to the mines, and even some 'rude shelters' built for the miners. The seam of aluminium sulphate at the far end of the mine was often less than an inch deep. Alum must have been of high value indeed if mining only became uneconomical when the seam was so small.[29]

The alum, as we have seen, was used in dyeing, tanning and in medicine, in Egypt and further afield. But the Egyptians themselves had come up with yet another use for this versatile resource: they found that if alum was dropped into cloudy water, the impurities coagulated on it, making it easier to filter them out. Thanks to alum therefore, Egyptians were probably the first people in the world to drink chemically treated clear water.[30]

Next to sweep through Egypt (in 332 BCE) was Alexander of Macedonia, and it was when one of his generals then seized power and reigned as pharaoh Ptolemy that a word and a concept emerge that will resonate in our story of alum: *monopoly*. Recognising that huge sums of money could be made from alum, Ptolemy ordained that from now on the state (i.e. he himself) would take control of its production and sales. Again and again, we shall find rulers in different ages and in different parts of the world passing laws to keep the income from alum for themselves.

The Romans did so, for one, when they occupied Egypt in 30 BCE. Indeed, they not only retained the Ptolemys' monopoly on alum, but maximised their income by militarising the oasis of Kharga, converting its temples into military garrisons and building forts along the major trade routes,[31] all in order to prevent the desert tribes from helping themselves to the alum as they had been doing before. The Romans kept a very close eye too, on both the contractors to whom the mines were leased and those responsible for transporting the alum to the ports. The state-paid officials who checked the loads on arrival were required to submit reports every five days – in sextuplicate![32]

By now Egyptian alum was facing foreign competition from mines in Spain, in Thrace (present-day Macedonia, Bulgaria and Romania), on the island of Lipari (north of Sicily) and on the Greek island of Milos. But sales of Egyptian alum soared all the same for the simple reason that it was head and shoulders above the others in terms of quality. Alum from Egypt and from Milos was, according to Aulus Cornelius Celsus in his book *De Medicina*, the most suitable for use in medicines. While Pliny the Elder, writing in the first century AD, rated Egyptian alum more highly than even that of Milos.

Pedanius Dioscorides, a contemporary of Pliny's and author of a five-volume treatise on medicine, advised using alum to treat gum disease.[33] And the Roman author Aulus Gellius then did his bit for alum in a story he wrote about how the Greek leader Mithridates, defending Piraeus against the Roman general Sulla, had coated his defensive wooden towers with alum. The result, according to Gellius, was that Sulla had been unable to set fire to the Greek fortifications, and his attack had failed. Here then, was yet another use of alum – as a fire-retardant. Exports of alum soon helped to make Alexandria one of the busiest ports in the Roman Empire and Egypt, Rome's wealthiest eastern province.

In 395 AD the Roman Empire broke into two entities and Egypt found itself in the eastern half, the part we have come to call the Byzantine Empire. From now on, therefore, the Byzantines were in control of Kharga alum, and all monies made from it accrued to their capital, Constantinople.[34]

In 552 AD, two monks smuggled silkworm eggs into Byzantium, breaking China's long-standing monopoly on silk production. The Byzantines now had access to Europe's highest quality alum and a monopoly on silk production within Europe,[35] and their silk factories began to churn out fabulously coloured silk clothes and furnishings, including brocaded silk, whose patterns were created by weaving contrasting luscious colours that were fixed using alum mordant. Such was Europe's insatiable appetite for coloured silk that Constantinople's export boom lasted fully six centuries, giving the Byzantine economy a massive trade surplus and huge gold reserves.[36]

And this fabulous wealth was crucial to the very survival of Byzantium, for as their once-vast empire withered and shrank, the Byzantines made a habit of buying off their enemies when they threatened Constantinople itself, rather than risking warfare. In 431 AD for example, when the Huns threatened Constantinople, the Byzantines promised to pay them 350 Roman pounds of gold annually if they went away. When the Huns reappeared in 434, the tribute was raised to 700 pounds, and in 447 the annual amount rose again to a massive 2100 pounds.[37] The wealth that derived from silk and alum played a large part in enabling Byzantium to survive a thousand years longer than the alum-less Roman Empire in the west.

Given that the two monks who reputedly brought the silkworm eggs from China were almost certainly Nestorians, it seems rather ungrateful of the Byzantines to have declared their leader, Archbishop Nestorius, a heretic[38]. Nevertheless, they did so in 431 AD, and they then banished him to their favoured place for exile in the desert – to Kharga.

Fast forward another ten generations and in the 7th century AD it was the turn of Muslim forces to conquer Egypt. During the brief period of confusion that ensued, the Bedouins once again helped themselves to alum from Kharga, carrying it by camel to Libyan ports such as Sirte, where they sold it, along with alum from the Kawar oasis in what is now Niger, to merchants from Venice, Pisa and Amalfi.

Egypt's Fatimid dynasty quickly restored order however, and set up a new state monopoly on alum run by an office called the *Matjar*. The *Matjar* bought the alum from Kharga and other sites in Egypt, supervised its transport down the Nile, fixed its sale price and oversaw the all-important export sales. A document from 1072 references a shipment of alum from Alexandria on its way to the Venetian trading post at Modon, in the Peloponnese, for onward transport to Venice.[39]

Indeed, Fatimid Egypt was happily trading alum with the Italian maritime states of Amalfi, Venice, Pisa and Genoa when, out of the blue, Christian warriors from Western Europe exploded into the region, ransacking cities and spreading terror, seizing Jerusalem in 1099, the port of Acre in 1104 and the port of Tyre in 1124. The crusades had begun, and alum was to play a pivotal part within them.

For despite the fighting, and despite the fanatical rhetoric against their Muslim foes, the truth was that Europeans still needed Egyptian alum, which was now supplemented by alum from Yemen. It was so vital to Europe's growing textile industries that Italian merchants continued to buy it from the Arabs to the tune of some 5,000 cantara[40] (about 300 tons) of alum a year. Throughout the crusades, therefore, Italian money filled Arab coffers.

The invaders marched into Egypt in 1163, the new Muslim leader, Saladin, pushed them out again in 1169, but the alum trade prospered despite it all, for we have records of Egyptian alum being delivered to the crusader-held port of Tyre in 1176, for example. And though there was a brief hiccough when Saladin regained Jerusalem and Acre in 1187, sales of alum soon soared again, hitting 13,000 cantara (1210 tons) in 1192. Ibn Mammati, one of Saladin's chief ministers at the time, tells us that much of this alum came from Upper Egypt, which included Kharga.[41]

The alum trade flourished despite the pope's repeated threats to excommunicate anyone who traded with the Muslim enemy. There were half-hearted bans on the trade from time to time (Venice claimed to impose one in 1224, for example)[42] but they were easily circumvented. So alum remained one of Egypt's chief exports all through the crusades, and indeed there were years in which sales of alum paid for fully *two-thirds* of all Egyptian imports – the vast majority of which were military-based, for they consisted of iron, weapons, wood for building ships, and even entire ships. Christian Europe's crying need for alum, in other words, was financing the purchase of the military hardware that the Arabs were using ever more effectively to defeat the Christian invaders.

The last act was played out in 1291, when the Mamluks (the Muslim rulers who had come to power in Egypt in 1250) drove the crusaders out of their very last stronghold of Acre. Decades of steady income from the exports of alum had made it possible to bring two hundred years of fighting to an end.

3

Alum, China and the First InfoTech Revolution

There's a revolution in information technology and China is at the heart of it.

Sounds familiar?

Like our own more recent revolution in digital technology, the world's first infotech revolution was triggered by inventions that dialled up the speed at which information could be exchanged. But while computers and the internet have impacted people across the world, the world's first infotech revolution was limited to just one country – China. This is where paper was first made, and where woodblock printing was born. And alum made a small but important contribution.

The very oldest piece of paper ever discovered is a fragment from about 150 BC found in a Chinese tomb,[1] and amazingly, even this early bit of paper has a scrawl of ink on it. We know therefore, that these pioneers faced a problem that still challenges paper-makers today: how to produce a surface that is strong and smooth enough to write on, porous enough to enable the ink to dry, yet not so porous as to allow the ink to smudge.

The world's first paper was made by breaking down the fibres in materials such as hemp, bark or even old rags, adding water, flattening the pulp and then drying it.[2] This process produced a surface that was stronger and smoother than that of natural bark, papyrus, bamboo or any of the other natural materials that had been used before. And the product was then further improved in a process called 'sizing', in which the paper was stretched taut and rubbed with starch. Sizing made the paper stronger, smoother, and less absorbent: any ink applied to it could now dry on its surface without smudging. As a result, paper was in widespread use in China by the third century AD.

Different centres of papermaking in China experimented with different processes, involving a variety of different ingredients, and in time *Xuan* paper made in Jing county in southeast China emerged as the very best. *Xuan* paper was made by pulping rice, bamboo and the bark of mulberry and elm in a process that could famously involve a hundred steps or more. It was so strong that you could write very legibly on it, and it did not deteriorate when stored. Not surprisingly, most of the early Chinese paper that has survived is *Xuan* paper.[3]

Even within *Xuan* paper however, there were three grades of quality, and the very highest grade of paper, called *Shu Xuan* paper, was sized yet again – and this time with alum. This additional sizing produced a surface that was superbly supple, yet so outstandingly strong as to be almost waterproof. And the wonderful strength of this higher-grade *Shu Xuan* paper sized with alum was crucial because it allowed China to capitalise on the second ground-breaking invention that revolutionised its information technology: woodblock printing.

Till the 8th century AD, every document, and every copy of every document, was copied laboriously by hand. This of course severely limited the number of documents and copies that could be made. It also resulted in copies that were littered with errors. The invention of woodblock printing in the course of the 8th century AD changed all that. Now, large numbers of copies could be made quickly, cheaply and accurately from one original, and what ensued was the same as when institutions in the 20th century installed their first printers: the demand for paper went through the roof.

There was a problem with woodblock printing though. To achieve the necessary clarity of script, hard wooden blocks had to be pressed repeatedly and extremely firmly against the sheets of paper, subjecting them to much rougher treatment than when they were written on by hand. Most grades of paper were simply not up to it: they just frayed and fell apart. But *Shu Xuan* paper had the necessary strength to stand up to woodblock printing, so the largest-scale dissemination of knowledge that the world had then ever seen (limited though it was to within China) was made possible only by paper sized with alum.[4]

And the consequences were massive. Duplication by woodblock printing enabled knowledge to spread throughout the country much as the internet has allowed knowledge to spread around the world. Centres of learning across China were soon inundated with block-printed documents – anything from religious texts and calendars to maths tables, almanacs and dictionaries.[5] There were even guides to the etiquette of weddings and funerals. And as the amount of information grew, administrators attempted to establish order in the chaos by imposing a system of categorisation. The eighth-century equivalents of our domain characters (.org, .edu, .com, .gov etc) were colour-coded sheets of paper: legal documents were on white paper, government business on yellow, religious communications on blue, and so forth.[6]

The widespread availability of printed materials on sturdy alum-sized paper raised China's literacy rates, allowing the country to make use of the brainpower of more of the population than had been possible before. The Chinese government could organise country-wide civil service exams in which young men (yes, men) from different backgrounds could qualify for government service, run as exams are run today: examiners were not allowed to know the identity of the candidate whose paper they were marking, and the final grade awarded depended on the judgment of two examiners who worked independently.[7] This was light-years ahead of Europe, where learning was restricted to a tiny number of monks and priests, and hand-copied books written on parchment were so rare that they were chained to the libraries' bookshelves. Public exams for the civil service would not appear in Europe till the 19th century! No wonder China remained for so long the most advanced and the most prosperous country in the world.

Of course, alum was not only used in the manufacture of paper. As elsewhere in the world, alum was also used in China in dyeing and tanning – indeed the role of alum in fixing colour was particularly important in China because certain colours were used to denote specific ranks. A practice that began with the wearing of coloured ribbons in the Qin Dynasty (221-207 BCE) grew in time into an elaborate system of coloured clothing, and by the time of the Ming Dynasty (1368-1644) higher officials wore red, middle-ranking officials wore dark blue, lesser officials wore green, while the common people were forbidden to wear any of these colours.[8]

Yellow was proof of especially high status, having been associated with the emperor since at least the 7th century AD. And two manuals from the 17th and 18th centuries tell us how the brilliant yellow was obtained. To produce a *bright yellow* you boiled a plant called 'pagoda bud' and soaked your yarn or cloth in the dye. To produce a *golden yellow* you added a plant called 'smoketree' to the mix. But whether *bright yellow* or *golden yellow*, the mordant to use was alum. Of course. For only alum could bind the colour firmly to the fabric and give the yellow its uniquely luminous shine.

No wonder then, that alum was highly prized by Chinese emperors. Alum was, in fact, one of the very few commodities over which they claimed a monopoly. Alum merchants had to obtain a licence, and as the emperor controlled the number of licences, he also effectively controlled the terms of trade. Over the years, of course, the licence fees went up, as did the proportion of the Chinese government's income that derived from alum.[9]

As in other countries too, alum was also used in medicine. China had its own sources of alum, but one of the grades of alum used in medicine during the Tang dynasty, which was very open to products and ideas from outside China, was called 'Persian white', clearly implying that it had been imported. However, Persia itself produced no alum,[10] so it is likely that Persia was simply the country of transit, and that 'Persian white' had originally come from Egypt. Could it even have been Nestorian Christians, who were active in China at this time, who brought the 'Persian white' alum to China? There is no evidence for this of course, but the Nestorians had experience of alum, their leader Archbishop Nestorius having spent years of his life exiled in Kharga.

Be that as it may, the use of alum in the manufacture of paper was a world first, and alum continued to be an ingredient in the production of the highest-quality printing-grade paper for an extraordinarily long period of time. Alum is there in the world's oldest printed book,[11] the so-called Diamond Sutra printed in China in the year 868, and there it is still in the Victoria and Albert Museum's collection of Chinese watercolours, painted in the eighteenth and nineteenth centuries on Xuan paper sized with alum and animal glue.[12]

A thousand years of use in making paper stronger and smoother. That was the enduring power of alum.

4

Alum in the Hands of Crooks

How much time did it take for crooks to realise that they could use a substance as versatile as alum for criminal purposes?

Shockingly little.

Evidence of fraud is, of course, notoriously hard to come by. By their very nature, fraudsters will do all they can to cover their tracks. And yet, astonishingly, we have hard evidence that as far back as Roman times crooks were using alum to make coins and jewellery of *debased* gold – that is, they were using alum to mix some silver or copper into objects that looked as if they were made of pure gold.

No Roman gold was one hundred percent pure. But Roman gold that has survived is of surprisingly high quality. Tests on gold jewellery manufactured in the first, second and into the third centuries AD from across the Roman Empire reveal an impressive gold content of about 23.5 carats (98% purity).[1]

Except in Egypt. Here, the same tests reveal that while the purity of the gold on the surface of items of jewellery is impressive, the gold content on the inside of artefacts is significantly lower. A gold bracelet with 99% purity on the surface, for example, shows only 94% on the inside. An earring with 97% gold content on the exterior has only 85% gold on the inside. And the gold content of the interior of a pendant with 90% gold on the outside is only 83.1%.[2] What was going on?

Clearly, craftsmen in Roman Egypt had mastered a technique which in the jargon is misleadingly called 'surface enrichment'. It might more honestly be called 'inner debasement', for the technique allowed you to produce items which retained the higher grade of gold on the outer surface, but substantially reduced the gold content on the inside. And we know precisely how the Egyptian craftsmen did it because, extraordinarily, two manuals from the third century AD have survived to this day. The manuals – papyri discovered in Thebes, on the Nile, in the 1820s, and later sold to libraries in Leiden and Stockholm[3] – contain 'recipes', or sets of instructions, for creating 'surface enriched' gold artefacts. And they have one thing in common. In each case one of the vital ingredients is alum.

Take Recipe 25 in the Leiden papyrus, for example:

> *1 Daub thin sheets of gold alloy with alum, salt and misy* [a mix of copper and iron sulphates]. *Heat the sheets and the base metals will separate out, leaving you with thin sheets of pure gold.*

So far, so similar to the Lydians' method of producing pure gold that we saw back in Chapter 2. It was the next step that was dodgy.

> *2 Apply your thin sheet of pure gold to the exterior of rings, pendants, bracelets and other trinkets made of debased gold. Your lower grade artefacts will now look as if they were made of pure gold.*[4]

It sounds easy enough, but of course coating items of jewellery with thin sheets of gold was a painstaking process requiring a goldsmith's expertise. So counterfeiters who had no experienced goldsmith among their number could turn to recipes that used simpler methods – this one for example:

> *1 Use Leiden Recipe 56 to manufacture your debased gold: simply melt four parts of gold with three parts of copper, and you will have turned a small quantity of 24-carat gold into a much larger quantity of debased 19 or 10 carat gold,*[5] *from which you can fashion your rings, bracelets, brooches, etc.*
>
> *2 Now use your alum to restore the gold content of the exterior. Bathe your object in a mixture of alum, salt and*

organic acid such as vinegar, and the base metal on the surface will leach out, leaving a surface of pure gold that can be polished up beautifully. The pure gold surface will suggest that the entire object is made of precious metal.[6] *Few customers will spot the difference.*

Or, alternatively, use Recipe 69:

1 Grind one part of alum and celandine with three parts of roasted misy, make into a paste with a child's urine.

2 Apply the paste to the surface of your gold object, heat and plunge into cold water.

Again, the base metal was removed from the outside, leaving a surface of pure gold. The anonymous author or authors of the papyrus guaranteed that the result would 'deceive even the artisans'.[7]

For those who could not afford to start off with even a small amount of gold, the papyri usefully provided recipes for *surface enrichment* treatment on silver instead. In fact, the Leiden papyrus alone has more than eighty (!) recipes for dissembling different metals, mostly gold and silver.[8]

But you did need to have access to alum: it was a key ingredient in just about all the surface enrichment recipes. That was no problem in Thebes, of course, with the alum mines of Kharga less than 300 kilometres away.

Now, coating a debased metal with pure gold was not, as such, illegal. The purpose was not necessarily fraudulent. Maybe goldsmiths in Roman Egypt thoughtfully produced lower-quality items of jewellery, labelled them as such, and kindly sold them at a lower price to customers who couldn't afford the real thing. Maybe.

But, at the very least, a trade in fake gold objects was now possible. And the Leiden and Stockholm papyri themselves suggest that fraud was rife because they include a recipe for testing whether a piece of jewellery was made of real gold or not. Melt the gold, the author writes, then reheat it and check its colour. If it has turned pale, it contains silver. Rough and hard? It has been contaminated with copper. Soft and blackened? A giveaway sign of lead. If however the gold retains the colour of a gold coin, you can breathe a sigh of relief: the gold in your trinket is pure.[9] There was a catch, of course. These tests transformed your treasured bracelet or pendant into a shapeless lump of gold. So counterfeiters will not, I imagine, have been too worried that they would be carried out too often.

Even stronger evidence of the fraudulent use of surface enrichment treatment lies in the progressive debasement in the gold content in Roman coins.

Unlike the coins of today, the value of Roman coins depended on the amount of silver or gold which they actually contained. For trade to function smoothly across the Roman Empire, it was essential that the coinage could be trusted from Hadrian's Wall to the Red Sea, and indeed beyond the confines of the empire itself, in places like southern India, for example, where some four thousand Roman coins have been unearthed, brought by Roman merchants who came to buy pearls and spices.[10] All such trade depended on buyers and sellers being able to trust the quality of the coins.

And tests on Roman coins have shown that while their quality was indeed maintained for centuries, their gold or silver content deteriorated significantly in the third century AD, the age in which the Leiden and Stockholm papyri were written. Silver coins suffered first. The 'silver' Roman denarius soon had only 12-18% silver content on the inside while retaining the silver-rich content on the outside.[11] And where silver led, gold followed: by the end of the third century AD the Roman gold *aureus* had become so thoroughly debased that, in 301 AD, Emperor Diocletian issued edicts against the counterfeiters. Moreover Diocletian, it seems, had a fair idea where the tampering with coins was going on because he tried to stamp it out by ordering the burning of all books about how to manufacture gold and silver – if they came from Egypt.[12]

Some hope! The genie was out of the box. With a little bit of gold, silver and copper – and, of course, a little bit of alum – it was now easy enough to counterfeit coins if you wanted to. You simply heated your gold or silver to melt into your copper, you struck your coin, and then you used your alum mixture to restore the gold or silver content of the surface.[13] If it was easy for criminal counterfeiters to debase the currency in this way, it was even easier for corrupt mint officials who had all the right equipment at their disposal. And it was easiest of all for governments to make cheap money by lowering the gold or silver content of coins in the official mints. No wonder the alum-fuelled practice spread like wildfire.

And not only within the Roman Empire. Aksum, a kingdom in present-day Ethiopia, Eritrea and Yemen that issued some of Africa's earliest coinage, had coins of a high standard till they began to be debased in the fifth century AD. Byzantine coins suffered the same fate, as is evidenced in an Egyptian document of 498 AD that records a loan of 'six *unalloyed* approved Imperial [Byzantine] solidi of gold'. The need to stipulate that the coins be 'unalloyed' strongly suggests that fake coins were in circulation.[14] And as far away as China we find a recipe in about 500 AD for using alum to make objects made of iron look as though they were made of copper. Although 'the nature of the interior remains', notes the recipe, 'the outside is copper-coloured'.[15]

Fast forward to the Middle Ages, and surface enrichment is still going strong. Twelfth-century coins excavated in Palestine have a silver content of 80% on the surface but only 60% inside.[16] In Venice, the medieval mint made coins from debased silver and then soaked them in a mix of alum, tartar, salt and urine to achieve a silver-rich exterior.[17] The fraud features in *The Evening and the Morning*, Ken Follett's novel set in England around the year 1000 AD[18] and in Europe's first book on metallurgy, *De la Pyrotechnia*, printed in Siena in 1540.[19] And 15th-century Scottish coins excavated in the Netherlands have a double connection with alum: many are crude imitations in which alum has been used in surface enrichment, and they had been brought there by Scottish merchants who were travelling in to the Netherlands to buy, among other things, alum.[20]

European migrants took the alum recipe for debasing coins with them when they crossed the Atlantic to the USA, with the result that counterfeiting was prevalent enough in Massachusetts for an early law to stipulate that a *coiner* could be 'fined ..., be set in the pillory ... and then have one of his ears cut off; and from thence be drawn to the gallows, and set thereon with a rope about his neck for the space of one hour, and ... be whipped ... and then be sentenced to hard labour for a term of not more than seven years.'

In vain. The USA was barely ten years old when a gang in Hampshire, Massachusetts, fraudulently diluted the silver content in the Spanish coins known as 'pieces of eight' (the coins most in use in the new republic, and legal tender in the USA till 1857). Exploiting the expertise of a skilled silversmith named Julius Frary, the coiners heated the silver coins with base metals to make poor-quality coins, and then enriched the surface by heating the fake coins in urine and a mix of chemicals. Which chemicals? You will have guessed it. Two of the gang's recipes have survived, and they both contain alum.[21]

The fact was that counterfeiting by surface enrichment was possible wherever alum and coins existed side by side, and that was all over the known world. So the crooks were busy till the value of a coin was no longer dependent on its metal content, but determined, as is the case today, by the figure embossed on its surface. By then the crooks had discovered a new way of making money from alum – in the adulteration of bread. We shall come to this later.

5

Alum, Venice and Genoa

In the history of the Mediterranean, Venice is the victor, Genoa the classic also-ran.

Who can fail to be seduced by *la Serenissima*, as Venice likes to call itself? The sensuous sweep of the *Canale Grande*. The swagger of the *gondolieri*. The stunning procession of *palazzi* with their suggestive Moorish arches. The stupendous arrogance of a city that flaunts four bronze horses stolen as booty from the Byzantine Empire on the cathedral roof of *San Marco*.[1] The intoxicating mix of ostentatious opulence, romantic decay and *louche* behaviour. The dazzle of an over-the-top Carnival and the annual spectacle of the Doges, now the mayors of Venice, wedding themselves to the sea itself. Not to mention such former Venetian possessions as Rovinj, Trogir, Dubrovnik, Korčula and Kotor, strung like pearls on a necklace down the Adriatic coast.

Poor Genoa, meanwhile, is lucky to make it into an online list of Italy's top ten cities. Tourists keep to the motorway that strides over the city on giant stilts – and even that collapsed one day in August 2018, hurling cars and lorries onto the houses below. Of the visitors who do venture into the city, many head straight for the ferry terminal to begin a Mediterranean cruise, the highlight of which might well be ... Venice.

And yet Genoa's churches and museums are also crammed with booty, and it too has *palazzi* built on the proceeds of gain and trade. For Genoa was for centuries a Mediterranean power that rivalled Venice, with overseas territories and trading posts of its own. Genoese ships plied the seas, making Genoa a lynchpin for trade in the Mediterranean and beyond. And at the heart of Genoa's phenomenal wealth was a key item of merchandise in whose trade the Genoese were even more successful than the Venetians.

And that item was alum.

⁂

Genoa's history, as the city emerged from the collapse of the Roman Empire, began inauspiciously. Stone watchtowers all the way up the coast of Italy are evidence that communities lived in fear of being raided and kidnapped by Muslim attackers, and when, in 935 AD, the lookout men in Genoa clearly failed to raise the alarm in time, the results were grim.[2] The marauders pillaged the city, killed anyone who got in their way, and sailed away with one thousand Genoese mothers, sisters and daughters taken prisoner to be sold on as slaves. A human disaster on an almost unimaginable scale. Genoa's rapid subsequent rise to power was fuelled, it is often said, by a determination that such a calamity should never ever be visited on Genoa again.

Power came from shipping. The Genoese built a fleet of ships and by 1000 AD they were trading alongside the established Italian maritime powers – Venice, Pisa, and Amalfi – in the lucrative markets of the Levant, as the eastern Mediterranean was then known. As early as 1064, when William the Conqueror's English secretary Ingulf was looking around for a vessel to bring him home after completing his pilgrimage to Jerusalem, the ship he found loading up in the harbour in Jaffa was Genoese.[3] And the Genoese were not picky. They happily made money from carrying Muslim pilgrims towards Mecca too.

Trade went hand in hand with fighting however. In 1016 the Genoese helped the Pisans evict a Muslim ruler from Sardinia, and the island later became Genoa's first overseas territory. And in 1087 the Genoese joined a Pisan and Amalfitan fleet of 400 galleys that ransacked Mahdia in Tunisia, in an attack that was as chillingly savage as the raid on Genoa in 935.[4]

Genoa was thus perfectly positioned to profiteer from the military campaigns that we have already seen from the Muslim perspective. Yes, it was the crusades that made the Genoese rich and powerful, not least by handing Genoa the keys of the alum trade to the west.

Were the crusades about recovering Jerusalem for Christianity? For some, yes, undoubtedly. But for the Christian Catholic west they were also a struggle for supremacy over the Christian Orthodox east. And for Italy's seafaring powers, they were a heaven-sent opportunity to make money. A lot of money, over two hundred years. Partly by charging fees for transporting the Christian armies, their food and their equipment, but more importantly by extracting concessions once victory had been secured: these might be sole trading rights in various precious commodities, or even better, preferential access to ports in the Levant. These ports yielded huge profits not only from produce from the Levant itself, but by giving access to goods from Persia, India and further east, without having to pay the taxes that Muslim Egypt imposed on goods imported via the Red Sea.[5]

For Italy's maritime powers then, the trick was to back the right horse – Frankish or Germanic, Catholic or Orthodox, Christian or Muslim. If your faction came out on top, you made a killing, with (for a while, at least) sole access to ports that teemed with goods from Syria, Persia, Egypt, Arabia, India and as far away as China. The military men you had helped to power kept your rivals out. Of course, back the wrong horse and you would find yourself excluded.

The First Crusade, launched in 1096, was an untold disaster for the inhabitants of Jerusalem, but a resounding success for Genoa. First, the city provided many of the 246 galleys needed to transport the troops and their equipment,[6] and then Genoa struck lucky in 1099 when some of its boats happened to sail into Jaffa (now Tel Aviv) just as the crusaders' attack on Jerusalem was beginning to stall. How can we help you? asked the Genoese. Wood! We need wood to build siege engines, the desperate crusaders replied. Wood? No problem, replied the Genoese and coolly began stripping the wood from their own ships. With this Genoese wood the siege engines were made, with the siege engines a breach was made, and through this breach Jerusalem was taken, resulting in a frenetic carnage in which neither man, woman, nor child were spared.

And the Genoese reaped their reward. The crusaders who they helped to victory awarded them buildings along entire streets in Jerusalem and Jaffa, and granted them lucrative trading rights in the key Mediterranean ports of Acre and Tyre.[7] And to show it was all really about faith, in case anyone should doubt it, the booty brought home to Genoa included such relics as the Virgin Mary's hair and a greenish vase-shaped vessel which the Genoese declared to be the Holy Grail. Round one to Genoa.

The seasoned old-timer Venice was of course none too pleased with this turn of events, and responded with a plan designed to knock out both its arch-rival Constantinople and the young upstart Genoa in one fell swoop: it coolly converted the Fourth Crusade, which set off in 1204 to attack Egypt, the greatest Muslim power in the Levant, into an utterly unprovoked attack on Constantinople, the greatest Christian power in the world. The city was ruthlessly sacked, hundreds of inhabitants were slain, the (Orthodox) Byzantine rulers were expelled, booty was seized (including the famous four bronze horses now in Venice) and the new Latin (Catholic) government showered privileges on Venice, while excluding Genoa. The Venetians followed this up by destroying 25 Genoese galleys in the war of Saint-Sabas in 1258, and driving the Genoese from their lucrative trading post in Acre.[8] Rounds two and three to Venice.

Genoa was left licking its wounds. So was Michael VIII Palaiologos, the expelled ruler of the Byzantine Empire, who had retreated to Nicaea, a Byzantine territory in northwest Turkey. Michael had troops but no ships. Genoa had ships and a burning need to take revenge on Venice. So when the two powers met at Nymphaion, near Smyrna (in Turkish: Kemalpaşa, near Izmir) in March 1261, a deal was struck: the Genoese would supply fifty ships to enable Michael to take Constantinople, in return for generous concessions if Michael became emperor.

It was a pivotal, all-or-nothing moment. The Genoese risked the wrath of the pope, who was furious with them for signing a treaty with his Orthodox rival. The Venetian-backed Catholic regime was well ensconced in Constantinople, protected by Venice's powerful fleet that had seen off attacks in 1235 and 1260. The chances of a successful campaign looked bleak indeed. But then, to everyone's surprise, when the ink on the treaty of Nymphaion had barely dried, and before the ships promised by the Genoese had even been delivered, a small number of Michael's soldiers were secretly let into Constantinople in July 1261 to find that the Venetian fleet was out marauding, leaving the city defenceless. The intruders joyfully opened the city gates, in poured Michael's Byzantine troops, out flew the Catholic occupiers, Franks, Germans, Venetians and all, and Michael VIII Palaiologos entered the city in triumph. The Byzantines were back in the saddle, and Genoa had backed the winning side after all. They had done precious little to make it all happen, but that did not deter them from queuing up for the concessions as promised in the treaty. And they were generous indeed.

The Genoese were allowed to return to Constantinople and establish a commercial quarter across the Golden Horn estuary in Pera (Galata). Genoa was also given the right to open trading houses on the islands of Chios and Lesbos, which later became colonies to rival the Venetian islands of Crete and Rhodes.

Above all, however, the Genoese were granted the port of Phocaea (present day Foça), on the western coast of Turkey south of Smyrna (Izmir) with its extraordinarily large reserves of good quality alum. For the Byzantine Emperor, it was a cheap way of keeping the Seljuk Turks and the Aegean pirates at bay: the Genoese would now do it for him.[9] And for Genoa, the alum deposits at Phocaea were like hitting the jackpot every day for two hundred years.

The Genoese were experienced in the alum trade.

In Africa, alum came from Taghaza, in the north of Mali,[10] and from the Kawar oasis, west of Lake Chad, where officials of the Kanem Empire guaranteed its safe transport on caravans of camels across the Sahara. Genoese merchants bought it at Sirte in Libya, Bugia (now Béjaia in Algeria), and other ports in North Africa. There are records of Genoese traders purchasing Egyptian alum in Alexandria in 1140, for example, and buying 250 cantara (12 tons) of alum and sugar in Ceuta in Morocco in 1191.[11] In Bugia a young Leonardo Bonacci, the son of a Pisan merchant and better known to us today as Fibonacci, realised that the numerals used by the Arab merchants were far more flexible than the cumbersome Roman numerals still being used by the Europeans. He decided to write a book to persuade Europe to adopt the Arab numerals, and when he seized on the calculation of the value of a ship's cargo as a practical instance to make his point, the cargo he chose, because he saw it being loaded all around him, was alum.[12] Alum thus played its modest part in converting Europe to Arab numerals when Leonardo's book, the *Liber Abaci*, appeared in 1202.

Genoa was also active in the alum trade in the eastern Mediterranean. The alum ordered by two Arras merchants from Genoa and forwarded to Bruges in 1225 may have been high-grade alum from Anatolia (present-day Turkey), Egypt or Milos (an island between Crete and Greece), or lower-quality alum from Spain, North Africa or Thrace (Macedonia, Bulgaria and Romania). But Genoa's activity in Turkey itself is confirmed by the Franciscan monk Willem van Ruisbroeck who in 1255 heard a Genoese trader in the Seljuk Turk capital Iconium (Konya) boast that Genoa and Venice between them had 'monopolised all the alum in Turkey'. The monopoly enabled the two rival maritime powers to hike the price of alum by two-thirds.[13]

On arrival in Genoa, some of the alum was sold to local Genoese and Lombard dyers, but most was sent north to Europe's largest cloth-making centre, Flanders. Six times a year,[14] caravans of mules laden with alum set off from Genoa, climbing step by step over the mountains behind the city. We think of Genoa as a maritime power, but the guild of muleteers was in fact one of Genoa's oldest,[15] and their caravans wended their way over the hills to the Aosta valley, then over the St Bernard Pass to the trade fairs in the Champagne region of northern France, Europe's most important place of exchange since at least 1127.[16] And here the Genoese traded their alum against cloth from the merchants of Bruges.

Of course, the overland route to northern France was slow and laborious for a bulky product like alum, so the Genoese had opened up an alternative route: some alum was shipped from Genoa to the French port of Aigues-Mortes, and taken north up the Rhone and Saone rivers.

This was faster, but also costlier because the dealers in Aigues-Mortes took their cut.

Why not then, we might ask, ship the alum directly to Bruges and avoid the fees of both Aigues-Mortes and the Champagne trade fairs altogether?

No doubt many a Genoese merchant had had this same idea. But it was impractical for a number of reasons. Partly it was the fear of the pirates in the western Mediterranean and the Muslim Almohads who controlled the shipping lanes between North Africa and El-Andalus (Muslim Spain). More importantly, galleys made money by stopping off and trading with ports along their route, and the ports on the west coast of France had little to offer the Genoese. The trial voyage of a Genoese galley to La Rochelle in 1232 was not repeated.[17]

Above all though, to make long-distance trading by ship profitable, you had to have a full boat-load of wares and Oriental goods such as pepper, spices, raisins and dates. Valuable though they were, they were not bulky enough to fully load a ship. Even the relatively small amounts of alum that made their way to Europe were not enough to fill a ship. It was cheaper to carry these goods overland, where money could be made by trading along the way.

But the port and alum mines of Phocaea, leased to the Genoese in 1261, turned all these calculations on their heads.

Phocaea was special in every way. The alum was high quality, the quantities vast, the alum mines were conveniently near a harbour that could take the largest boats of the age, and the two Genoese brothers who took over the management of the mines, Benedetto and Manuelle Zaccaria, were larger-than-life characters whose energy, business acumen and forward-thinking transformed both the alum business and the Mediterranean trade routes.

First the Zaccarias radically raised production. They expanded the workforce, so that by 1307 they were employing upward of three thousand men and women in Phocaea's alum industry.[18] But the miners did not extract the natural alum that we saw back in Kharga[19]: in Phocaea they mined a whitish, pinkish or yellowish rock called alunite. And alunite had to be subjected to a series of processes to yield pure alum.

So there were workers who roasted (or *calcined*) the rock in large kilns or in open-air fires. The burning might take several days, or indeed several weeks, depending on the rock. More workers then piled the burnt rocks into giant stacks which they sprayed with water until the rock fragmented into rubble, a process that could take anything from five weeks to three or four *months*. The decomposed ore was then shovelled into huge basins and steeped in water, following which more workers again collected the resulting liquid and boiled it to obtain crystals of water-soluble alum.[20]

Phocaea was, clearly, a full-scale industrial site. Production levels soared, and when the Florentine merchant Francesco Balducci Pegolotti visited Phocaea in about 1340 he calculated that it was churning out fourteen thousand cantara (670 tons) of alum annually, more than any other alum mine he had visited in the Levant.[21]

Each step in this complex process involved cost, and the Zaccarias also bore the cost of keeping the local pirates under some measure of control. The Zaccarias' manufactured alum was therefore more expensive than natural alum. But it was also purer, because the burning and crystallising processes freed the alum from impurities. When used in dyeing, therefore, alum from Phocaea produced colours that were even more vivid than had ever been achieved with natural alum. If the Zaccaria brothers, then, could get their alum to Bruges, the dyers would happily pay over the odds for the extra quality it gave to their cloth.[22]

To begin with, the Zaccarias sold their alum to Genoese traders who took it to Genoa's newly acquired trading station on the island of Chios. Here it was stored along with mastic, pepper and other spices till galleys called to ship it to Constantinople, Cyprus, and on to Genoa.

But the canny brothers soon realised that they could make more money if they cut out the middlemen, so they acquired a fleet of ships and shipped their alum to Genoa themselves. They then opened their own dyeing business in Bisagno, outside Genoa, thus creating what has been called one of the world's first vertically integrated commodity empires, for they were now involved in the production, distribution and exploitation of their alum.[23] No family had ever dominated the trade in any single product to such an extent before.

And the Zaccaria brothers had a crucial advantage as far as shipping was concerned for, virtually uniquely in the eastern Mediterranean, they had a heavy commodity with which they could bulk out their ships. By filling their holds with alum, and then topping up with expensive pepper, spices, wheat, fruits, weapons and cotton, they could fully load their ships, reduce their payload and turn in a tidier profit. Which explains why in 1277, only two years after taking over the mines of Phocaea, the Zaccarias were able to do the unthinkable. They filled a galley with alum and other saleable goods, and sent it not only to Genoa but – risking the pirate threat – out into the Atlantic, across the Bay of Biscay and all the way to the northern European harbour of Bruges.[24]

Just imagine the sensation when the Zaccarias' galley, with its crew of over one hundred oarsmen working two sets of oars, pulled into the harbour in Bruges. A ship from fabled Byzantium had magically, yet literally, landed in the heart of Flanders! Its load of something over twenty tons[25] of merchandise might seem paltry to us, but for the citizens of Bruges, the galley's holds held more exotic goods than any mule train had ever been able to carry by the overland route.

The arrival of the Zaccarias' alum galley in Bruges, one might well say, marked the moment that seaborne trade moved out of the Mediterranean and into the Atlantic,[26] and the moment that Christian Europe's[27] three most advanced urban societies – Constantinople, the city-states of Italy and the cities of Flanders – first came into direct trading contact with each other. The consequences were not only an increase in trade, but an exciting, fertile inter-exchange of ideas in art, fashion, culture, philosophy and politics which, it can be said, launched the European revival and set the stage for Europe's domination of the world.

That, however, was for the future. For now, the Zaccarias were laughing all the way to the bank. First they made a fortune by selling their alum and other goods. Then they bought enough high-quality Flanders cloth to fill their galley once more, and made a second killing when they sold this Flemish cloth in Genoa and in Constantinople. The new trade route – risky though it was in terms of pirates and the danger of shipwreck – was win-win for the Zaccarias, for Bruges and for Genoa.

And, very soon, for England too. England's sheep produced the best wool in Europe, and while much of it was exported to Flanders, England had enough of a cloth trade to make it the second largest market for alum in Europe after Flanders. It didn't take long then, for the Zaccarias to tap into this market too: their first galley to England arrived in Southampton the following year (1278),[28] to the same furore as their first galley in Bruges. And it returned packed with English wool for the cloth makers of Genoa.

With the Genoese now able to transport larger volumes of alum and with savings of up to 33% as compared to the land route,[29] age-old patterns of trade were utterly transformed. The cloth makers of Bruges, with more high-quality alum at their disposal, were able to produce even better-quality coloured cloth. And Flanders cloth and English wool now flooded Mediterranean markets. Florentine merchants who had for centuries come to Genoa to buy Spanish wool, were by the end of the 14th century buying English wool there instead.[30]

By 1300 Genoa was on a roll.

In 1284 the Genoese had decisively beaten the Pisan fleet near the island of Meloria, off the Tuscan coast, and had then eliminated them as rivals by building a wall at the mouth of the river Arno, silting up their harbour. In 1291 and 1292 warships sent to the western Mediterranean by the Zaccarias quelled the pirates and defeated the Muslim fleet which had control of the waters between Africa and Spain. By 1295 the Genoese had the largest number of galleys the city would ever have in its long history: 165 of them – of which only eighty were needed to inflict a painful defeat on Venice in 1298 off the Croatian island of Korčula.[31]

The money made by Genoese merchants enabled them to build magnificent palaces and contribute staggering amounts of money to Genoa's coffers. By 1293 the taxes from Genoa's seaborne trade suggest a taxable income of nearly four million Genoese pounds. To put it another way, the single city-state of Genoa, with a population of maybe 100,000, had something like *ten times* the receipts that accrued to the French royal treasury in 1292.[32] Read that statistic again. It really is extraordinary.

Not all this wealth, of course, was made from alum. Genoese merchants traded a wide range of products with Constantinople, Chios, and Egypt, Muslim and Christian Spain, and North Africa. Gold from south of the Sahara, for example, allowed Genoa to become the first European power to issue gold coins in 1252. But the trade with Bruges was a key element in Genoa's wealth, because alum – high-grade alum from Phocaea, supplemented by lower-grade alum from Spain, North Africa and Lipari – formed the bulk of the cargo of the Genoese ships going north to Bruges and Southampton, with wool taking its place on the return trips.[33]

Alum, then, gave Genoa the edge over Venice in Flanders and England. The first Genoese galley had arrived in Bruges in 1277; the first Venetian galley would not arrive till 1314.

Roll over, *Serenissima*.

The alum trade justified Genoa's own nickname: *La Superba*.

6

Alum in Bruges and Southampton

It's a rare thing in Europe, but the tallest and most exquisitely chiselled medieval towers in western Belgium and northern France are, in the main, not those of churches or cathedrals. Instead, they grace cloth halls, guildhalls and town halls, and they exude civic pride rather than religious fervour. More than fifty (!) of them are listed as UNESCO World Heritage sites,[1] and as befits a city which was for almost two hundred years one of the largest and most powerful in Europe, the 83-metre belfry of the cloth hall in Bruges, built in 1280, is – in my opinion at least – the queen of them all. A symbol of the city's status, the belfry tower also protected the city's wealth. Climb its stone steps and you'll come to the original 700-year-old wrought iron gates that guarded the city's treasury, its coffers amply filled by income derived from the city's cloth industry.

Alum arrived in Flanders about a hundred years before this belfry tower was built.[2] It came overland, mainly via Arras, at that time the largest and busiest of the Flanders cloth towns. This was largely because Arras was the closest Flemish town to the all-important Champagne trade fairs where, in the twelfth century, goods from northern Europe were exchanged for goods carried overland from the south. What raised Bruges to its predominant position in Flanders was the seaborne galley trade,[3] kickstarted, as we have seen, by the arrival of the Zaccarias' alum-filled galley from Phocaea in 1277. From the very start then, alum was at the heart of the triumph of Bruges.

And what a triumph! Thanks to its direct connections to Genoa and Constantinople, Bruges rose in the fourteenth century to four times the size of London.[4] Its wealthy citizens wore pearls, jewels, precious stones and Chinese silk; Turkish ceramics and Persian carpets were abundant, as were exotic peppers and spices, figs, dates, raisins and other Mediterranean fruits, all brought by Venetian, Florentine and Genoese galleys and later, the larger, square-rigged caracks.[5] A crane powered by men working a treadmill unloaded goods from the Baltic, the Atlantic and the Mediterranean. Elegant four-, even five-storeyed brick-built buildings lined the streets. The trading houses of the Italian and German states, Spain, the Hanseatic League and rich banking families like the Medici and the Fugger all had trading houses in Bruges, which doubled as the most significant money market in northern Europe.[6] As J. A. Van Houtte puts it, 'Nowhere was there outside Italy an area so densely populated, where a numerous and wealthy *bourgeoisie* provided so considerable a purchasing power and so profitable an outlet.'[7] The city's wealthy merchants and their wives were some of the best-dressed people in Europe.

Alum was crucial to Bruges as an essential ingredient in the industry that produced most of this wealth: the manufacture of high-quality coloured woollen cloth. The city's weavers worked with English or Scottish wool, the best in Europe, and its dyers used top-grade dyes, such as kermes, brazilwood and lapis lazuli. But they needed alum both to fix the colours in the cloth and to give the luminous sheen that we see in the luxurious gowns, cloaks, dresses and even headwear so lovingly painted by Jan van Eyck, Hans Memling and Rogier van der Weyden.[8] And Bruges enjoyed a key privilege: it was the alum staple for the Burgundian Netherlands, which meant that all the alum imported into Flanders had to be channelled through Bruges, where the appropriate customs duties were paid and collected.

Top quality Turkish alum was imported throughout Bruges' golden age, much of it unloaded down river in Damme or Sluis, as Bruges' river, the Zwin, gradually silted up. The guilds maintained the highest standards of quality by carrying out spot checks to make sure that no mordant other than high-quality alum was used in Flanders cloth. In some cities, such as in Ypres, even the measure of alum was stipulated.[9]

While the dyers toiled out of sight in smelly workshops filled with harmful acidic fumes, the alum merchants in Bruges were soon wealthy, well-known figures who operated from offices that exuded luxury. By the 1400s the city's alum trade was largely in the hands of the Genoese

Lomellini family, who organised every stage of its transport from Turkey and brought 60,000 cantara (288 tons) of alum in 1416 and 1417 alone. Filippo Lomellini had it loaded into ships in far-away Pera (the Genoese commercial quarter next to Constantinople), Joffredo Lomellini oversaw the transit of goods in Genoa, while Georgio Lomellini organised the unloading in Bruges. Georgio then sold the alum to the dyers in Bruges, Ghent, Ypres and other weaving towns in Flanders, or traded it on to Bremen, Hamburg and Hanseatic Baltic ports. The sizeable profits were used to buy good quality cloth which Georgio then shipped back to Joffredo and Philippo.

But the prosperity could not, of course, last for ever. Bruges imported 50,000 cantara of alum in 1445, but only 90,000 cantara during the following three years.[10] The numbers were falling, the glory years of Bruges were coming to an end, and the cause for the city's decline lay in England.

The weavers in Bruges relied on English wool, but from the 1300s the English government had begun to discourage the export of 'raw', unworked wool in order to encourage the growth of a home-grown cloth industry. High customs duties on the export of wool caused the exports of English wool to dwindle from 35,000 sacks a year in the 1350s, to 14,000 sacks in the 1420s, and a mere 4,000 sacks by 1500; while, in contrast, exports of English *cloth* rose in leaps and bounds from 6,000 pieces of broadcloth annually in the 1350s, to 39,000 pieces in the 1420s, and 82,000 by 1500. What's more, while English cloth had previously been largely uncoloured, leaving the weavers and dyers of Bruges masters of the trade in coloured cloth, by the end of the fourteenth century English weavers and dyers had caught up with their Flemish rivals and almost all the English cloth that was exported was coloured.[11]

The dearth of unworked English wool and the rising quality of English-coloured cloth spelt the doom of the Flemish cloth industry. English weavers had an in-built advantage over their Flemish competitors: they had their own local supplies of high-quality wool. Flanders, in contrast, had no wool of its own, so Flemish weavers were reliant on imports. They tried to replace imports of English wool with Spanish wool instead, but this lowered the quality of their product and further undermined their trade. Cloth production in Ypres fell by more than 70% by the end of the fourteenth century.[12]

In desperation, Bruges banned the import of English cloth in an attempt to protect the jobs of its own weavers, but it was too late: the English cloth trade simply diverted to Antwerp, and took the Italian and German merchants with it. Trade in Bruges withered, the dredging of the Zwin was abandoned, and when angry unemployed artisans rose up against the sovereign ruler, Maximilian of Austria, he retaliated by ordering all remaining foreign merchants out of Bruges in 1488.

The very final indignity came in 1491 when Maximilian stripped Bruges of its status as the Netherlands' alum staple, and transferred that privilege to Antwerp.[13] Alum had launched the golden age of Bruges, and alum marked its end.

Bruges now entered a period of decline, its population rapidly falling to a quarter of what it had been a hundred years earlier, and Antwerp, the new alum staple, became one of the richest cities in the world.[14]

In Southampton, as in Bruges, the arrival of the Zaccarias' first galley initiated two hundred and fifty years of trade based on wool, cloth and alum.

True, the evidence of Southampton's trade with the Mediterranean is not written as gloriously over the cityscape as it is in Bruges, but traces survive in unexpected places – in the church of St Nicholas, North Stoneham, for example, (now next to Southampton airport), where the medieval Italian text on a limestone slab tells how it once covered a vault in which Italian-speaking oarsmen were buried far from their homeland.[15]

The Genoese, and later the Venetians and Florentines, came to England for one reason only: to buy English wool, widely regarded as the best in Europe. The country had anything up to nine million sheep in the early 1300s,[16] at a time when historians suggest a human population of under three million. Many were in seriously large flocks belonging to the richer monasteries. Southampton offered the double advantage of a large harbour close to the sheep-grazing hills of southern England, and – for the times, at least – good road connections to London (avoiding the pirate-infested sea route), where the Italians could sell their alum, wine, fruit and wax. Southampton therefore served the Italians as a convenient outport for the capital, much as Sluis served Bruges.[17] In addition, Southampton offered access to England's southern market towns: deliveries of alum, madder, raisins, soap, almonds and wax are recorded arriving by packhorse and by wagon in Bristol, for example.[18]

The king welcomed the Italian merchants, and happily granted them legal protection, recognising that duties on their trade would significantly boost his revenues. Parliament, however, which was dominated by the mercantile classes who saw the Italians as unwelcome competition, tried to throw spokes in their wheels. The status and security of the Italian merchants see-sawed, therefore, according to whether the monarchy or Parliament was top dog. In 1335 and 1351 Edward III was strong enough to decree that foreign merchants be allowed to trade without fear of discrimination; but in 1376 Parliament banned foreign merchants from selling directly to the public, and compelled them to appoint an English agent to 'host' their business.[19] And in London, they were forbidden to reside for more than forty days at a time.

Indeed, from time to time mere mistrust of foreigners spilt into outright anger and violence, and this was another, darker reason why the Italians preferred Southampton to London. In Southampton, where few local people were engaged in commerce, the Italians were not seen as rivals and were not subjected to as much abuse as in London.[20]

However, the outbreak of war with France in 1337 made the Genoese vulnerable even in Southampton, because Genoa allied itself with France. In October 1338 a force of several thousand French and Genoese men attacked Southampton, setting houses alight, plundering the warehouses and killing many inhabitants and taking others to be sold into slavery. This, of course, exposed the Genoese alum traders to reprisals.[21] As late as 1389, two English ships attacked a Genoese trading ship, forced it into Southampton and seized its cargo on the grounds that the Genoese were an enemy power, and when the Genoese merchants appealed, the mayor and bailiffs of Southampton sided with the attackers. The Genoese cargo was saved only by direct intervention of the king.[22]

Fortunately, though, such occurrences were the exception rather than the rule. In Southampton, relations between the foreign merchants and the local population were friendly on the whole, and by 1377 the town had the third highest number of foreigners – known as aliens (!) – of any town in England, the largest group among them being Italians.[23] The Genoese worshipped at the Friary, the Florentines at St John's, and the Venetians at the Chapel of St Nicholas.[24] Damiano de Pezaro, a Venetian who lived in Southampton for fifteen years, became a freeman of the city, and Christopher Ambruogi (or Christopher Ambrose, as he was known locally), a Florentine who traded in alum, wine, fruit and confectionery, was mayor of Southampton from 1486 to 1487 and again from 1497 to 1498.[25] There were also some marriages of Italian merchants to English women.[26]

However, as Southampton had no local cloth industry, the wealth brought by the Italians did not percolate through the local population as it did in Bruges. Local men were employed in the more menial tasks of loading and unloading the ships, while the more lucrative jobs, such as trading alum with dyers in the cloth towns, were in the hands of merchants from London, Salisbury and Coventry. Despite its alum trade, therefore, Southampton remained a poor city. Even its richer traders, such as Walter Fetplace who imported dyes and mordants (i.e. alum), and Robert Bagworth who owned a ship and traded in cloth, came nowhere near rivalling the merchants of Bruges.[27]

The disparity in wealth between the two cities is almost painfully evident in the few medieval buildings that have survived in Southampton. The 12th-century merchant's house now known as Canute's Palace, the 13th-century weigh house and the 14th-century wool warehouse (significantly known as the Alum Cellar) are each of them sturdy enough stone-built buildings, but coarse and unsophisticated in comparison with their equivalents in Bruges. And just compare Southampton's simple half-timbered Woollen Cloth Hall with the majestic Cloth Hall in Bruges! The buildings make it clear that Southampton was the *second* most important destination for Mediterranean alum.

On the other hand, the alum trade in Southampton reached its peak in the mid-fifteenth century, just as Bruges was declining. The two trends were linked, of course, because the boom in English-coloured cloth that spelt the downfall of Bruges caused English dyers to require more alum than ever. As a result, in the years 1438-39 alum and dyes for England's cloth industry accounted for fully one-quarter of all the goods imported into the country,[28] and by early 1466, 3,880 cantara (roughly 185 tons) of alum, worth £4,187 (approaching three million pounds at today's prices), were being imported into Southampton,[29] and some dyers in London were now wealthy men.[30]

Paradoxically, however, the decline of Bruges ultimately had a negative impact on Southampton, for the switch in trade from Bruges to Antwerp created a new trading axis between London and Antwerp in which Southampton was bypassed entirely. The city's role as the gateway to London disappeared, and the city's alum trade had collapsed by 1490, and had disappeared almost entirely in the sixteenth century.[31] With the death of the alum trade, there was no longer any reason for the Italian galleys to call, and the last one left in May 1532 – some 250 years after the Zaccarias' first alum galley had triumphantly entered Southampton Water.

And with no more Italians, no more galleys and no more alum, Southampton, like Bruges, slid into a long period of decay.

7

Alum, Slavery and the Black Death

Crenellated walls loop around the tops of rocky headlands and race up craggy mountainsides. Turreted fortifications line the shore, their massive stones reflected in the waters of the Black Sea. At Enisala on Romania's Black Sea coast, a gaunt empty-windowed tower raises its head to the sky like an eyeless giant. At Sudak in the Crimea, a fourteen-towered wall protects the mighty keep. At Amasra on the northern, Black Sea coast of Turkey, the Cross of St George, the patron saint of Genoa, is carved into the castle wall. And indeed all around the coast of the Black Sea, though many were built by the Romans and extended by the Ottomans, the castles are invariably called Genoese. Six of them are listed as UNESCO World Heritage sites.

What's going on? We are not on a crusader route. We are in a region populated by Greeks, Turks, Russians, Byzantines, Mongols, Tatars, Jews, Persians, Ukrainians, Georgians, Armenians, Bulgarians and Romanians. And we are some 1900 miles by sea from Genoa.

Genoese? Why Genoese?

We have to go back to the Treaty of Nymphaion of 1261, which granted Genoa generous concessions in return for helping to oust the Venetians from Constantinople and crown Michael as Byzantine emperor. When Michael became emperor Michael VIII Palaiologos later the same year, he awarded Genoa the commercial quarter of Pera, the islands of Lesbos and Chios, and the all-important alum mines in Phocaea. But that wasn't all. During the Venice-backed Latin occupation of Constantinople, Venetian ships had been allowed to sail from the Mediterranean, through the Bosporus (that vital narrow sea waterway that is not much wider than a river at the point where it passes Constantinople) and into the Black Sea. The Venetians had been given a monopoly on the trade with ports on the north Anatolian (Turkish) and Crimean coasts, while Genoese ships had been shut out. But after Nymphaion, the situation was reversed. Michael closed the Bosporus to Venice and opened it to Genoa.

Whooping with triumph and lured by the prospect of silk, spices and jewels from the legendary Silk Road, the Genoese sailed their ships north into the Black Sea, turned east along the north Turkish coast and found ... alum.

High-quality alum. Better quality even than the Zaccarias' alum from Phocaea.[1] In limited quantities however, because unlike in Phocaea the alum was mined not on the coast, but 120 kilometres inland at Koloneia in the Seljuk Turk territory of Etena.[2] Back in 1246, the Dominican friar Simon de Saint-Quentin had compared the value of these alum reserves to that of a silver mine,[3] and in the mid-14th century the widely-travelled Florentine merchant Pegolotti would reckon that there was as much alum at Koloneia as in Phocaea. The Seljuks used much of this alum themselves in the production of their famous Anatolian carpets,[4] and they controlled its trade to Persia, to Aleppo, to the crusader towns and Cyprus, and north to the Black Sea coast. So the Genoese had to make do with as much alum as the Seljuks chose to send on caravans of mules or camels on the seven-day journey over the Pontic Mountains.[5] No matter! The Genoese recognised top-quality alum when they saw it. They knew that alum from Koloneia would fetch a wonderfully high price in Bruges and in England and were soon happily shipping it through the Bosporus and on to Genoa.

Back in Phocaea, however, the Zaccarias were not amused. The new trade in Black Sea alum undermined their near monopoly which had made them immensely rich, and they had no intention of accepting this unwelcome competition without a fight. They were perfectly capable of putting together a fleet of warships, but on this occasion, they hit on a neater, cheaper and sneakier plan. The alum from Koloneia had to be shipped through the Bosporus past Constantinople. Theirs did not. Very well, then. If they could persuade Emperor Michael VIII Palaiologos to ban the export of alum through the Bosporus, the competition would be choked off at a stroke.

The emperor owed the Zaccaria brothers a favour because they were subduing the Aegean pirates at no cost to the emperor himself. What's more, Benedetto Zaccaria was maybe even married to one of Michael's daughters (he certainly called his son Paleologo, not exactly a common name). So a word in the emperor's ear was easily had, a handsome gift was probably made, and the desired effect was certainly achieved, and in 1275 Michael VIII forbade the export of alum from the Black Sea through the Bosporus. The Zaccarias rubbed their hands in delight. Their business was secure. The Black Sea alum trade was scuppered.

Or was it? Back in Genoa, the merchants were up in arms, furious that the Byzantine Emperor had gone back on his promise to grant them free access to and from the Black Sea and livid at being deprived of a lucrative new source of income. Genoa would not take this lying down. Michael had to be shown who held the whip hand. The blockade must be broken. So early in 1276, a galley set sail from Genoa with instructions to teach Michael a lesson. It sailed demonstratively through the Bosporus past Constantinople, while the Byzantines looked on helplessly.[6] It then loaded with Black Sea alum, and prepared for the journey home.

But this time the Byzantines were waiting for it. They captured the ship, brought it ashore, and then upped the ante by confiscating its cargo of alum and punishing the crew in the traditional Byzantine manner. Every member of the crew was blinded.[7]

This of course enraged the Genoese, who now resorted to their nuclear weapon: they ordered their vessels to stop supplying Constantinople with grain from Anatolia and Thrace, or with any other foods on which the city depended.[8] A food boycott. What had begun as an inter-Genoese spat about alum had spiralled into the largest city in the Eastern Mediterranean being threatened with starvation.

And this, let's remember, was the behaviour of two 'friendly' Christian powers. Constantinople and Genoa were actually allies. Trading in this region was not for the faint-hearted. Powers traded with each other, betrayed each other, helped each other and knifed each other in the back in bewildering twists and turns. When they quarrelled, some people were hurt, some were killed, some sold into slavery and some made a fortune. And then the *status quo ante* was restored. In this case, the ban on trading Black Sea alum through the Bosporus was lifted in 1278.

I wonder what the men who had been blinded made of it all.

The quarrel with Michael had shown the Genoese that they needed a secure base in the Black Sea region. And they obtained one from what might seem an unlikely new ally – the Mongol successors to Genghis Khan. After he had swept westwards three decades or so previously, Genghis Khan's vast empire had fractured into four (still vast) self-governing Mongol 'khanates', and one of these, named the Golden Horde, whose rulers had converted to Islam[9], had recently conquered most of Crimea. And these rulers from the Far East were more than happy to sell one of their Crimean ports to the newcomers who brought such wonderful cloth, iron, glassware, wine, oil and weapons from the Far West. Late in the thirteenth century, therefore, Genoa acquired the Crimean port of Caffa (now Feodosia). And Caffa was a game-changer.

In Caffa the Genoese had finally got what they had so nearly achieved in Acre before the Venetians turfed them out in 1258: their own trading post at the western end of the fabled Silk Road trade route from Persia, India and China. Here they could buy pepper, ginger and other spices, Chinese silk and Indian cotton, pearls, jewels and exotic wood directly from the Mongols – who charged export duties of 3-5 percent, far lower than the fees and taxes of up to 20-30 percent charged by the Mamluks in Alexandria.[10] What's more, the entire route to Caffa from the east was under the control of Mongol khanates, so although the caravans of camels took nine months or more to make the journey, the route from China to Caffa was, according to our ubiquitous friend Pegolotti, 'perfectly safe, whether by day or by night' – certainly safer than the sea route on which merchants risked drowning in a sudden monsoon storm, or being attacked by pirates and sold into slavery.[11]

However, the goods had to be sent on by boat to Genoa, and here, as we know, you made money by filling every nook and cranny of your boats. You could not do this with the high-value, lightweight goodies from the east, so this is where alum came into the story.[12] Bulk out your boat with alum, and a tidy profit was assured.

Caffa therefore began to import alum from around the Black Sea, some for its own use, but mainly to be re-exported as ballast in the ships and as a lucrative product to sell at the far end. Five hundred cantara of alum from Trebizond in Anatolia arrived in Caffa in 1289,[13] and more arrived the following year from both Trebizond and Sinope, a Seljuk port a little further west .[14] In both cases, the alum will originally have been from Koloneia.

Even the Venetians bought alum in Caffa, for Michael soon allowed Venice back into the Black Sea as a counterweight to the Genoese. While never trading as much alum as the Genoese, Venice carried enough to supply itself and southern German cities such as Augsburg and Nuremberg. As a result, Caffa exported a massive 9,000 cantara (nearly 430 tonnes) of Anatolian alum in 1289-1290, even after keeping 57 cantara (2.7 tons) of alum for its own use.[15] Alum helped to make Caffa profitable.

Caffa was very different from the Genoese trading posts in Pera, Lesbos and Chios, which were commercial quarters in Byzantine territories. Caffa, instead, was a thriving polyglot city in its own right, peopled with Greeks, Russians, Jews, Armenians, Poles, Tatars, Mongols, Turks, Persians and more. Estimates of its population range from 20,000 to 70,000 and more, making it comparable to a larger European city (London for example had 60,000 to 70,000).[16] It had a castle, and its two sets of concentric walls were over five kilometres long.[17] It had shipbuilding yards,[18] and acquired a court, a mint, and its own bishop. This was no mere trading post: this was a fully-fledged colony,[19] funded by the customs duties charged on each vessel that used its harbour.

But although they ran the place, the Genoese were always a minority in Caffa, so a rough and ready tolerance of religious and cultural differences existed here, unheard of in the heartlands of Europe.[20] Caffa had Catholic, Orthodox and Armenian Christian churches as well as a mosque and two synagogues.[21] This open-to-all attitude sat alongside a rampant, no-holds-barred money-making ethos that attracted any young man eager to risk his all for fortune. So Caffa was at once secure and precarious, brimming with opportunity yet dangerous. Every trading ship from Genoa had to make a perilous, 60-day journey across the pirate-infested Mediterranean, through the dangerous straits of Messina and Gallipoli, and through the Bosporus which the Byzantines could close at any moment. The risk of shipwreck was great, both at sea and within harbours: a Genoese ship sank in the harbour of Genoa in 1347 and another in 1350.[22] But for those who made the trip successfully, the profits made on every successful voyage were swash-bucklingly, money-spinningly brilliant, and Caffa was soon the Black Sea's central *entrepot* for trade in every imaginable commodity.

So Caffa rapidly became Genoa's 'Wild East', the place to make a fast buck. Not, you understand, the sort of place to which you would bring your wife and children, as Genoese merchants did in respectable Pera and Chios.[23] Life here appealed to men who, weary of a monotonous existence back home, were excited by the thrill of an unscrupulous city in which morality played second fiddle to making money, where religious, national or family allegiances and conventions were flouted in the pursuit of naked profit. Small wonder, then, that alum and grain only occupied slots two and three in the export commodity rankings.[24] In male-dominated, money-driven, winner-takes-it-all Caffa, the chief export was not horses, nor timber, dye, furs, leather, soap, wax, honey, salt, fish, nor even wine or caviar, but live human beings.[25]

Slavery in the Eastern Mediterranean was as old as the Bible. Slaves were depicted in Egyptian hieroglyphs, Assyrian bas-reliefs and Greek vases. Persian, Roman, Mongol, Byzantine and Seljuk Turk armies routinely sold captured peoples into slavery. Indeed, all free persons in the Mediterranean lived in the knowledge that they themselves could one day be captured and enslaved, as a result of a coastal raid, of war or of shipwreck. Genoa itself had, as we know, been sacked in a raid. Now the boot was on the other foot, and Genoa was fully involved in capturing people and enslaving them. They seized twenty people from their homes on the Bulgarian coast in 1352, for example,[26] and carried away 5000 people to be sold as slaves after sacking Alexandria in 1365.[27] Let that figure sink in. Five thousand human beings, trafficked and sold.

Nor were only coastal communities at risk of being sold into slavery. People were seized far inland too, especially in the steppes of what is now Ukraine and Russia, from where they were marched to Caffa. The Genoese alum trader Francesco Draperio, who would control the alum mines of Phocaea from 1445, sold more than 500 slaves in five years between 1445 and 1451.[28]

Slavery was indeed accepted as a normal, legal fact of life, and permitted under Catholic Christian, Orthodox Christian and Muslim law.[29] When in 1246 the pope condemned the Genoese, Venetian and Pisan merchants for transporting slaves to Egypt,[30] it was not on the grounds that the trade was immoral, but that Christian states should not be helping a Muslim foe. And when Pope Gregory famously saw a child for sale in a slave market in Italy as early as 590 AD, he was moved by the fact that such an angelic child should come from a land that was heathen, not by the fact that the child had been enslaved.

And that is important because the Black Sea slave trade was largely in children. The median age of the seventy people sold into slavery by the notary Lamberto de Sambuceto in Caffa in 1289-1290 was twelve. The median age of the 251 enslaved people sold by a different notary in 1359-1363 was fourteen.[31] The average age of human beings at their first point of sale was 11.3 years.[32] And even though there was a special trade in young light-skinned males to serve in the army and civil service of the Mamluk rulers of Egypt, the majority of human beings who were sold were female. Of the slaves whose gender and age are documented, 80% were female slaves aged between 8 and 24.[33] Sexual abuse was widespread and horrible.

The trade in live human beings pre-dated the Genoese in Caffa, but it was during their occupation that the port became the largest market in enslaved people in the region.[34] Sardinia, Spain, the Balkans and Africa all supplied slaves to the Mediterranean, but the number of slaves from the Black Sea outstripped them all.[35] Every ethnic group in Caffa was involved. Tatars rode out into the steppe and captured innocent children.

Russians and Mongols sold prisoners taken in battle. Armenians, Greeks (Byzantines), Jews and Genoese acted as brokers. Genoese, Venetian, Byzantine, Seljuk, Mamluk and Catalan traders carried slaves from Caffa and sold them in Cyprus, in Egypt, and in Europe.

People were captured for nothing but sold for a high price, so the profit margin was enormous. In terms of money, the 114 slaves aboard a ship from Caffa in 1455 were 'worth' almost double the value of the ship itself.[36] Young slaves were sold in Genoa at the end of the 13th century for the cost of six oxen, the annual rent for six store rooms, ten pieces of woollen cloth, or three years of a sailor's salary.[37] So Genoese citizens in Chios had slaves, rich Egyptian households had slaves, and over four thousand citizens of Genoa – notaries, silk workers, wool and cloth merchants, barbers, priests, monks, nuns and doctors – owned one or more slaves, making the unfree population almost 10% of Genoa's population in the early 1400s.[38] This was typical of northern Italian cities.[39]

The alum and slave trades converged in the men who bought, sold and transported what they saw as two commodities. For there were no specialised slave ships or slave merchants in Caffa. The merchants who inspected, purchased and shipped alum were the same men who, when human beings were marched into Caffa with ropes around their necks, did not feel pity but saw a business opportunity. The slaves were paraded naked so that the merchants could inspect them,[40] a contract was drawn up and signed as for any other wares, and the men, women and, more often than not, children were then loaded, often in small numbers – usually less than twenty, rarely more than forty[41] – onto a boat along with the grain, the furs, and the alum that the ship might also be carrying. The enslaved human beings were then carried along with the other goods to Egypt, to Genoa, to Majorca, to Cadiz and, yes, to London, Southampton and Bruges.[42]

Somewhat to my surprise, the oarsmen who powered the galleys that carried the alum were free men, as the practice of using galley slaves apparently only began in the sixteenth century.[43] But while slaves were largely in domestic service in Italy, male slaves were routinely exploited for hard labour in the eastern Mediterranean, in the sugar fields of Crete,

Cyprus and Syria, for example. So were slaves used in the alum mines in Phocaea and elsewhere? Most material I have read about Phocaea is strangely silent on the subject. However, a contract of 1446 shows that the Duke of Naxos supplied both free and enslaved workers for the Venetian-controlled alum works on the island of Milos,[44] and a few sources do state explicitly that slaves were worked in the alum mines of Phocaea.[45] No wonder the profits were good.

By the 1340s the Genoese were enjoying the fruits of success all round the Black Sea, and had settlements from Kilia and Likostomo in the Danube delta to the Gonio fortress in what is now Georgia.

Caffa bought its alum in Trebizond, a Byzantine territory on the northern Turkish coast that was independent of Constantinople, and in Sinope, which was under Seljuk Turk control. And the city also traded with an Italian trading post which was, incredibly, even further to the east: this was Tana (present-day Azov) on the river Don in Russia. Life here was even riskier than in Caffa, for Tana could only be reached by negotiating the narrow straits of Kerch, crossing the Sea of Azov and sailing up the river Don.[46] But with Tana being so far east, the Silk Road was here at its very shortest, which explains why both Venetian and Genoese merchants felt it was worth the risk.[47]

However, in 1343 a tiff between a (Catholic) Italian and a (Muslim) Mongol in Tana spiralled out of control when a Muslim man was killed and Mongol soldiers under their leader Janibeg threatened the Italian settlements. The Venetians and Genoese merchants promptly abandoned Tana and retreated to Caffa, but Janibeg and his soldiers followed hot on their heels and laid siege to Caffa through the winter and into 1344. They could not, however, besiege the seaward side of the city, so the alum ships from Trebizond and Sinope could bring in food and supplies. Then, when a relief party arrived from Pera, the Genoese broke out of the city, destroyed the Mongol siege machines and killed a few thousand Mongol soldiers. One estimate of the collateral damage of this particular dispute was 15,000 dead.[48]

But that was not the end of the affair. Janibeg and his troops were back in 1346. Again they besieged the city, again they harried the Caffans from the landward side, and again supplies were brought in by sea. But this time the Mongols were more tenacious, and the inhabitants of Caffa were under constant fire from rocks and stones catapulted over the walls by the Mongol *trebuchets* – catapults that could hurl weights of up to a hundred kilos.

The situation was deteriorating fast when in May 1347, suddenly, inexplicably, all enemy activity ceased.

Was this a trap? The Caffans peeped nervously over their walls, and could hardly believe their eyes: all around were dead bodies stacked in piles. The Mongols had been struck by a killer disease that had decimated their forces. Victory, thought the Caffans, was theirs. Proof positive that a Christian God had saved them from their Muslim enemy.

But while the disease had fatally weakened the Mongols, it had also provided them with a weapon. The *trebuchets* were worked again, and this time it was not rocks and stones that rained on Caffa but disease-infected corpses. Hundreds, maybe thousands of them were catapulted by the Mongols over the walls and into the city. So many that the Caffans could not clear them all away, and the rotting corpses infected the air and poisoned the water. Soon, inevitably, Caffans themselves began to show signs of sickness.

Such is the tale as reported in 1348 by Gabriel de Mussis, a notary in Piazensa in Italy. Did the catapulting of infected corpses really happen? No, say those historians who suspect that the story had grown in the telling. Yes, say others, pointing out that medieval catapults were quite capable of doing the job, and that the Mongols, who believed that the disease emanated from dead bodies, will have wanted the corpses out of the way as quickly as possible. Some claim this episode was maybe the world's first case of biological warfare.[49] But it is also possible that the disease was transmitted by rats passing between the besiegers and the besieged. Or by sailors on the alum ships from Trebizond that were supplying the city with food. Or indeed by slaves from further east who were brought to Caffa on those same ships.

But whatever the causes of the disease, the consequences were dire. In fear for their lives, the Genoese abandoned Caffa in droves, leaving the indigenous population to its fate. First they sought refuge in Pera, hoping to return to Caffa when the plague was over, but when the pestilence hit Pera at the end of June, they fled west again along the alum route to Venice and Genoa. Alum had made this trade route profitable, but it now exacted a terrible price, with Genoa and Venice the first two cities in western Europe to succumb to the plague. It was 1347 and the Black Death had arrived.

And, of course, it spread. From Venice, lice, rather than rats and fleas, carried it along the alum trade route to Austria and southern Germany. But the alum sea route from Genoa to England was faster, so the Black Death struck Southampton, Bristol and London even before it reached Augsburg, Nuremberg or Vienna.

The Black Death cost maybe 25 million lives in Europe and devastated swathes of Asia, the Near East and North Africa. Maybe 200 million human beings died in all.

Yet ironically, Caffa and Tana, which had played a key part in the devastating chain of events, emerged virtually unscathed. The dispute in Tana was patched up, the Venetians and Genoese returned, and life and trade appeared to go on as before.

But appearances were deceptive. The chaos caused by the Black Death contributed to the implosion of the Golden Horde and, when fighting broke out, the northern Silk Road that led to Crimea became unsafe.[50] Traders then switched back to the southern route to Egypt, Caffa's long-distance trade dried up, and the Genoese were forced, once again, to buy their silk, spices and jewels in Alexandria, at prices dictated by the Mamluks.

Strangely though, Caffa did not wither. On the contrary, Caffa boomed and Genoa went on a building spree.[51] The Galata Tower in Pera was built, the Yoros fortress extended where the Bosporus meets the Black Sea,[52] and Caffa's own citadel reinforced with walls two metres wide, eleven metres high, and over seven hundred metres long.[53] And Genoa was active militarily, too. It seized Cembalo (Balaklava) from the Principality of Theodoro (a Christian dependency of Byzantine Trebizond) and Soldaya (Sudak) from the Golden Horde. When Genoa then took Vosporo (Kerch) at the eastern end of Crimea, its territory extended one hundred kilometres along the Crimean coast, with control of the Sea of Azov and the route to Tana.[54]

How was this all made possible in the absence of the fabled wares from China and India?

The answer is that Caffa grew rich on the more mundane local produce from within the Black Sea area itself: wheat, cereals and fish to feed Constantinople and Trebizond, slaves, leather, wax, salt, fruit, furs – and of course alum.[55] There was high-quality alum from Koloneia, delivered to the Genoese commercial quarters in Trebizond and Sinope, and from at least 1395 also a lower-grade alum from Kütahya, in western Anatolia.[56] Incredibly, there was even alum brought by camel to Tana, probably on a 1,000 kilometre trek from the Ganja mines in Azerbaijan.[57] That such an enormous journey could be contemplated is proof of the high value of the product.

Moreover, when ships sailed from Caffa, they often sold some of their high-priced luxury goods in Pera and topped up with even more alum before heading on to Cyprus, Genoa, Bruges and Southampton. There was plenty on offer: alum from Ulubat at ports in the Sea of Marmara, alum from Kütahya at the Aegean ports of Altoluogo (Ephesus) and Palatia, alum from Phocaea, now run by an association of Genoese families called the 'Mahona', and alum from the mines on Lesbos managed by the Genoese Gattilusio family.[58] Two thousand and five hundred cantara (119 tons) of alum were loaded onto a ship here in just three days in 1397.

The largest ever recorded tonnage of alum was 18,000 cantara (857 tons) of alum in 1395.[59] This was made possible because the Genoese replaced their 300-ton galleys with sail-powered cogs and carracks with 600, 700 and, by the early 1400s, up to 950 tons of storage space. It was, in fact, the alum trade that had driven this development in ship design because in the first half of the fifteenth century, as Charles Verlinden puts it, the Genoese fleet was essentially an alum fleet, with alum the most important item transported by Genoese ships returning from the east, whether measured by weight or by value.[60]

However, while alum ensured that the Genoese flourished in the Black Sea region for another century, it also financed the rise of a rival power that would one day send the Genoese packing.

When in 1352 the Ottomans signed their first trade deal with a western state, the state was Genoa, and the commodity for which the Ottomans exacted a handsome sum was alum.[61]

8

Alum and the Ottomans

1453 is one of those rare pivotal dates in history. Like, say, 1066 in England, it is a date that has not just been seized on as a convenient watershed by historians, but was experienced by people living at the time as a dividing line between two eras.

For the fall of Christian Constantinople to the Muslim Ottomans in 1453 was not just the surrender of a significant city. It was the end of the eastern Roman Empire, a thousand years after the western half had collapsed. It was the end of the Byzantine Empire that had once included Egypt, Mesopotamia, Turkey and the Balkans and the end of Christianity's claim to supremacy in the Eastern Mediterranean. Even when the Byzantine empire had shrivelled to a ghost of its former self, the impregnable city at its heart, Constantinople, exuded stability and commanded respect with its large population, ancient ceremonies and massive protective walls. The city's incredible riches, to which sales of alum made a significant contribution, allowed it to buy off its attackers. Forget Rome: Christian Constantinople was the real Eternal City. It had been there for eleven hundred years. In human terms, that means it had always been there. It was a point of reference for friend and foe alike. When it fell, the world changed.

Constantinople had been on its knees before, but had always rebounded. So why was 1453 different? How did the Ottomans achieve their victory?

Without giving it much thought, I had vaguely imagined the Ottomans sweeping through Turkey in an unstoppable onslaught, much as Genghis Khan and his Mongol hordes had swept through Asia before. In my mind, the Ottomans had taken Constantinople as the culmination of a dramatic lunge westwards, and the victory had then opened the door to further Ottoman conquests in Europe.

But it was not like that at all.

In fact, a Muslim power had existed in the region well before the Ottomans. They were the Seljuk Turks, who defeated the Byzantines in 1071[1] and built a vast empire stretching from India through Persia to Turkey (Anatolia), at much the same time as William the Conqueror was building his much smaller one in Normandy and England. The Seljuks occupied all of Turkey except the Trebizond Byzantine mini-Empire on the Black Sea coast, and tiny enclaves in the west – such as Phocaea – that still belonged to Constantinople. From very early on then, the Byzantines had eyed a rival, dominant Muslim empire across the sea.

Battered by the Mongols, the Seljuk Empire had fragmented in the thirteenth century, to be replaced by a dozen or so semi-independent Muslim Emirates that jostled for supremacy. The Ottomans were one of these and easily the most successful. Their first leader, Osman I,[2] extended his territory within northwest Turkey. Orhan seized territory in Europe from the Byzantines as early as 1352. Murad I seized more land in Europe and his new capital, Edirne, was a European city well to the west of Constantinople. So no: the fall of Constantinople did not open the gates to the Ottoman conquests in Europe. It was the other way round. Ottoman conquests in Europe had already fatally weakened Constantinople. And no: the Ottomans did not sweep in from the east. They had already cut off Constantinople from the west.

Alum comes into the story when we consider how these decades of war were financed. Conquering powers often fund their military exploits by imposing heavy taxes on their newly conquered peoples, but the Ottomans famously taxed their conquered citizens very lightly. Instead they paid for their troops and military hardware by very savvily controlling the trade in two key commodities, grain and alum, and keeping their prices at the upper end of what the market would pay.

The beauty of it from the Ottomans' point of view was that their principal customers were their Christian enemies – the Byzantines, the Venetians and the Genoese – who, in the absence of any alternative sources, were dependent on the Ottomans for these products. The more the Ottomans raised the price of grain and alum, the more the Christian powers funded the very military campaigns that would ultimately be aimed against them.

The Ottomans had dominated the alum trade well before they took Constantinople. In 1326 they took the alum mines of Ulubat in north-west Turkey, along with Triglia, the port on the Sea of Marmara from which Ulubat's alum was exported.[3] In 1361, they seized the alum mines of Sapes in what is now north-east Greece, though they went on being managed by the Gattilusio family, who also ran the mines at Mytilini on Lesbos.[4] And in 1389-1390 they took control of the alum operation at Kütahya[5] along with its export ports of Altoluogo, Palatia, Antalya and Castro.[6]

The Ottomans suffered a brief setback when the resurgent Mongols under their leader Timur[7] defeated them at the Battle of Ankara in 1402, but the Ottomans were back in control of most of Turkey by the 1420s. As in Sapes, the Ottomans were perfectly happy to farm out the running of the alum mines and the management of the alum ports to the Genoese – for a fee, of course.[8] Indeed, given that the Genoese often collected the fees and taxes charged in the alum ports, they had become, in Kate Fleet's words, 'functionaries of the state', indeed 'an integral part of the embryonic Ottoman state.' And all the while they were providing the Ottomans with a steady, guaranteed income which could be put to military purposes.[9]

And Phocaea? Phocaea continued as a tiny Byzantine enclave that clung very vulnerably onto the western coast of Turkey. Its alum mines were owned by the *Mahona* a society of powerful Genoese families,[10] and run by Genoese 'farmers' (the term used for those who actually ran the enterprise). However, the Genoese realised that although Phocaea was still under the rule of the Byzantine Emperor, their alum operation now *de facto* existed at the mercy of the Ottomans, who could invade at any time. The massive stocks of alum stored at their base in Chios, estimated at some 100,000 cantara (4760 tons) in 1424, were indeed a very tempting prize for any potential invader. So the Genoese responded in a time-honoured manner: they sweetened relations with their new neighbours by paying their respects and agreeing to pay them an annual sum of 2,000 ducats.[11] Not a bribe, you understand, still less a tribute. Let's call it an annual gift. And the Ottomans graciously accepted it. Why not? This arrangement enabled them to make money even from alum mines that were *not* under their control.[12] What's more, the payments would give them leverage over Phocaea if ever the Genoese were unable to pay.

Which is exactly what happened next. When wars in the Mediterranean disrupted alum sales, Phocaea's income dropped, and by 1421 the *Mahona* was six years in arrears with its payments to the Ottomans. Would the newly crowned Ottoman sultan, Murad II, use this as a pretext for invading Phocaea and taking its lucrative alum mines?

Luckily for the Genoese, Murad II had an even grander goal in mind, for in 1422 he began the first Ottoman siege of Constantinople. For three solid months, Ottoman cannonballs pounded the city walls while siege machines flung heavy boulders at the city walls. Things were indeed looking bleak for the Byzantines when, out of the blue, Murad received news which put his own position in jeopardy: his brother Mustafa had raised an army, launched a revolt in mainland Turkey, and was now challenging Murad for the leadership of the Ottomans – just as Murad and his whole army were on the European side of the Bosporus, without a fleet to get them home. If Murad could not get his soldiers back over to Turkey quickly, the rebellion had a real chance of succeeding.

Enter the Genoese, sniffing a heaven-sent opportunity to save their alum mines. 'Might Murad wish to loan a few Genoese ships?' they asked. 'Genoese ships could be put at His Lordship disposal. If, of course, His Lordship were willing to overlook the little matter of Phocaea's debt...'[13] The deal was done, Genoese ships ferried the Ottoman army over to Turkey, Mustafa's rebellion was put down, Murad survived as Sultan, and the rest is history.

Could Mustafa's rebellion have fatally weakened the Ottomans if the Genoese had not supplied the ships? There is of course no way of knowing. But civil wars within the ruling family did at different times fatally weaken the Mongol, Byzantine and Trebizond empires. And what is certain is that if not his brother, certainly no other power in the region was in a position to challenge Murad's advance. The Pope's appeals for intervention by European powers fell on deaf ears. A half-hearted Venetian attempt to build an anti-Ottoman alliance came to nothing. Following Murad's aborted siege of Constantinople, the Byzantine Emperor himself acknowledged the reversal in the balance of power and began to pay the Ottomans an annual tribute.[14]

So it can be said that by putting their desire to save their Phocaea alum mines above all else, the Genoese scuppered the possibility of an Ottoman civil war, the one eventuality that might have postponed or prevented the Ottoman conquest of the region.

With hindsight, the behaviour of the Christian players in the eastern Mediterranean seems extraordinarily short-sighted. How could they so utterly underestimate the power of the Ottomans?

Bear in mind that the Christian powers had seen the Mongols, the Seljuk Turks and other powers come and go. They had grown used to dealing with shifting balances of power, and had come to believe that they could always buy their way out of any short-term crisis. So to the end, instead of taking on the Ottomans, the Genoese, the Venetians, the tiny Byzantine Principality of Trebizond and the even tinier Byzantine Principality of Theodoro in the Crimea continued to score points off each other. In 1423 the Theodorans attacked the Genoese trading city of Cembalo, and the Caffans spent huge sums defending the city. In 1431 the Venetians went to war (again) against Genoa. In 1433 a Theodoran-backed rebellion took Cembalo from the Caffans, following which the Genoese recaptured Cembalo and destroyed the Theodoran port of Kalamita. As late as 1446 thirteen galleys from Trebizond took part in a raid on Caffa.[15] The Christian powers were well and truly doing the Ottomans' work for them, mutually weakening each other, unwilling or unable to see that the Ottomans were playing the long game.[16]

The alum merchants were as blind to the tidal wave that was about to hit them as the military leaders. Phocaea was booming in the hands of a wealthy and ambitious Genoese merchant called Francesco Draperio, with 15,800 cantara (745 tons) of alum mined and exported in 1451.[17] And Draperio had eliminated some of the competition by agreeing to pay the Gattilusios 5000 ducats[18] a year to cease alum production in Lesbos. The deal was struck in 1450 and the duration was set at ten years.[19] Ten years! Some hope!

Only the Byzantine Emperor John VIII, back in Constantinople, saw the writing on the wall. He pleaded with the kings of Poland and Hungary to come to the rescue of Christianity in the east, and even offered Rome the union of the Catholic and Orthodox Churches, under the Pope's authority, in return for military help. But this offer only angered John's people back home who saw it as an act of humiliation. A last 'crusade', now no longer to capture Jerusalem but simply to defend Constantinople, came overland only as far as Varna in Bulgaria, where it was roundly defeated by Murad II in 1444. There would be no help from the west.

So when the final dramatic blow came in April 1453, Constantinople stood virtually alone. The sum total of Christian aid to this most significant city in Christendom consisted of 700 Genoese soldiers, five Venetian and five Genoese vessels, and a handful of Catalans. This time the Ottomans had a fleet of their own, and pounded the ancient city walls with cannon balls up to 60 cm in diameter and up to 600 kilos in weight. In 1422 the city had successfully withstood three months of siege; in 1453 it collapsed in half that time. The walls were breached, the Ottomans streamed in, and by the end of May, Sultan Mehmet II had declared the city to be his new Ottoman and Muslim capital of Istanbul.[20]

The Genoese in their walled quarter of Pera had grandstand views of the siege. These citizens had of course sworn allegiance to the Byzantine Emperor but, maybe sensibly, though not exactly bravely, they decided that they had met their obligations with a financial donation in 1452. Once the siege began, they lifted not a finger to help their Byzantine lord and neighbour, but instead looked on to see how events would unfold. Then, once the outcome was clear, they surrendered the keys of Pera to the victorious Sultan. This may not have saved their honour, but it saved their city, for the Sultan rewarded them with a charter guaranteeing their right to property, livelihood, religion and trade, and indeed awarding them a degree of self-rule unmatched anywhere else in the Ottoman Empire. It came at a price, of course: like other non-Muslims in the empire the inhabitants of Pera were now subject to a poll tax.[21]

Pera, then, had survived. Despite their victory, the Ottomans, it appeared, were prepared to strike deals with surviving Christian enclaves. This was reassuring. There was maybe no need to panic after all. If the Genoese could negotiate terms for the survival of Pera, could they not do the same for Caffa and their other possessions? Might this tide them over till the Ottomans were evicted from Constantinople? Many Christians in the region were sure that the Muslim conquest of Constantinople could only be of short duration.[22]

But anyone who had not yet realised that the world had changed for good was in for a rude shock.

Mehmet II needed to show that he was now master of the whole region, and where better to assert his authority than in the alum trade? First Mehmet tripled the tribute paid by the *Mahona* in the port of Chios, where the alum was stored and shipped, from 2,000 to 6,000 ducats. Then, in 1455, he sent a fleet to seize Phocaea.

And who was on board – some say even the leader of the fleet – but Francesco Draperio! Yes, the head of the alum mines under Byzantine rule had done what wily magnates do when confronted with regime change: he had quickly re-invented himself as a friend of the new regime, and had offered to lead the fight against his former colleagues in Phocaea. No doubt Draperio hoped his manoeuvre would be rewarded with the right to continue running the lucrative alum mines, but he was left empty-handed. For when the Ottomans seized Phocaea, they decided to manage the alum mines themselves, even if this led to a temporary decrease in production while the new managers and workforce learned the necessary expertise. And later they farmed out the alum mines of Phocaea to a Venetian consortium.[23]

With Phocaea now under Ottoman control, the only Aegean territories still in Genoese hands were the islands of Lesbos, with its alum mines, and Chios, with its alum-filled warehouses. Many Genoese hoped that these reserves would supply the market until Christian forces regained the upper hand. And indeed, the alum stocks on Chios were so enormous that the *Mahona* continued to feed the Genoese alum trade for another five years! In 1455 nine large alum ships carried some 71,000 cantara (3500 tonnes) of alum from Chios to Flanders and England,[24] and in 1456 a ship loaded up with 5,500 cantara for England, with a further 1,000 cantara to be dropped off in Cadiz.

Ironically, one reason the supplies lasted so long was that the alum mines on Lesbos had re-opened. Three years before, Draperio had tried to close them. Now they were vital to maintaining supplies. That's how the world had changed.

All of which suited Mehmet very well: he simply raised the tribute to be paid by the *Mahona* in Chios from 6,000 to 10,000 ducats in 1456, and 30,000 ducats in 1457 – a fifteenfold rise on the amount paid only five years before.[25] The alum trade that had helped to fund the taking of Constantinople was continuing to contribute handsomely to the Ottoman war chest.

And of course, the alum stocks in Chios could not last forever. From year to year, they were diminishing along with any prospect of Christians re-conquering Constantinople. And as the supply of alum dwindled, its price rocketed. In Genoa, it rose from about 2½ lira per cantara before 1453, to 6½ lira in 1455, and between 8 and 10 lira in 1458. In Venice, a merchant noted that the price of alum had increased ninefold by 1462.[26] Moreover, as the stock of alum reduced on Chios, Italian traders were forced to buy more and more Ottoman alum from Phocaea, Kütahya and Sapes – all at the new high prices. Mehmet celebrated by seizing the very last Genoese-run alum mines on Lesbos in 1462, cutting 400 Italians in half and enslaving many of the others.[27]

While the high cost of alum was a headache for the merchants back in Genoa, it was welcomed by Genoese who lived some three thousand kilometres further east – in Caffa.

Yes, Caffa. For, astonishingly, even though the Ottomans now controlled both sides of the Bosporus, the Genoese still hung on in their Black Sea colony. True, the Caffans now paid tribute money to the Sultan (first 2,000, then 3,000, and then 4,000 ducats a year)[28] and were supposed to submit their ships for inspection in Istanbul (Constantinople), where they had to pay both customs charges and protection money. Even then, some Genoese ships were shot at while in Istanbul, so many Caffa-bound ships decided instead to run the gauntlet and sail through the Bosporus in defiance of Ottoman regulations. Trading in and with Caffa, in other words, had become even more dangerous, even more macho.

Not surprisingly, an increasing number of Genoese appointed as administrators in Caffa were politely turning down the job offer.[29] But for merchants still willing to take the risk, the profit margins in Caffa were now even more lip-smackingly enormous than before, as the sale price of commodities from the Black Sea had soared. Alum from Koloneia, which the Caffans could buy in the still-existing Byzantine territory of Trebizond, could now be sold in Genoa for four or five times as much as before.

The high price of alum made it worth opening an overland trade route to Lwow in Poland (now Lviv in Ukraine), where some alum was used in the Polish cloth industry, while the rest was sent westwards to the German lands.[30] But the main trade route remained through the Bosporus, and the blockade-busting gunrunners of the American Civil War were nothing against the alum runners of Caffa, who dodged cannon balls from castles on both sides of the Bosporus, not least from the recently strengthened Rumeli Hisary fortress. Picture it: the Bosporus is seldom more than three kilometres wide, and only 750 metres from shore to shore at its narrowest point. With cannons firing from both sides, it is astonishing that any Genoese ship got through at all.

That Caffa held out as a Genoese colony in this unbelievably perilous way for another twenty-two years speaks volumes about the tenacity, bravery, not to say sheer foolhardiness of its Genoese merchants. But of course it could not last. Mehmet was mopping up any enclaves that did not yet belong to him, taking the tiny Trebizond Byzantine Empire in 1461, Tana in 1471,[31] and the high-quality alum mines of Koloneia in 1473.[32] These had been the last alum mines in the Black Sea area not under Ottoman control, and Koloneia now took on its Turkish name of Sebinkarahisar (from the Turkish şap = alum, kara = black, and hisar = fort). The Christian workforce was retained, however, in return for which they were spared the *devshirme,* or child levy – the terrible system by which the Ottomans forcibly took male children from Christian families to be converted to Islam, and then trained as soldiers and court officials. [33] The alum mines at Sebinkarahisar were in fact worked by Orthodox Greeks till they closed at the beginning of the 20th century.[34]

Caffa's own turn came in 1475. Its 70,000 inhabitants resisted for seven days, but then the Ottomans marched in and put an end to Genoese rule.[35] Three thousand slaves working in Caffa – men, women and children – were taken from their Christian masters, and the Muslims among them were freed. The Genoese, along with the Venetians who had fled to Caffa from Tana, were deported by ship to Pera, after handing over half the value of their possessions in cash. But not all. Those aged 20 or under, maybe between 1500 and 5000 Italian boys and 450 Italian girls, were made slaves, along with Poles, Russians, Wallachians, Georgians and other inhabitants. Only the Greek and Armenian populations were spared.[36] The Genoese colonial project had begun in 935AD to avenge the kidnapping of a thousand Genoese women; it ended five hundred years later with over a thousand Genoese young women and men, boys and girls, being led away into slavery.

The harsh reality of the new world order had finally sunk in. The Ottomans were here to stay. Constantinople was lost to Christendom for good. The Byzantines were finished, and so were the Genoese possessions in the Black Sea.[37] And alum production was now entirely in Muslim hands: the Mamluks controlled Egyptian and Yemeni alum, the Ottomans controlled the rest.

Genoese, Venetian, Catalan and Florentine merchants therefore had no option but to ignore the pope's injunctions and buy their alum from Muslim countries.[38] Not only the aristocracy but an expanding *bourgeoisie* in Europe, especially in Italy and Flanders, had developed a taste for luminous alum-mordanted cloth for their garments and soft furnishings, so while the price of alum was sky high, so too was the demand. So alum remained the last commodity shipped in bulk from the Muslim east to the Christian west.[39]

The cost to Christian Europe was enormous. The Genoese alum merchant Giovanni de Castro estimated in 1462 that the alum trade was bringing the Ottomans 300,000 ducats a year,[40] and it funded the Ottoman armies that marched to the gates of Belgrade in 1456, and took Athens in 1458, the Peloponnese (southern Greece) in 1460 and Bosnia in 1463.

On taking Constantinople, Mehmet had boasted that it would not be many years before he conquered Rome itself. And it was now looking very much as though the money being paid for Turkish alum would enable him to do just that.

9

Glut and Crisis in Italy

Hold your nose!

You're in late medieval Italy, you're about to visit dyers in their workshops, and I should warn you: it won't be a pleasant experience. Yes, I know, we all love the sumptuous colours that dyers create in woollen cloth, but the processes they use to obtain these colours ... well, to put it bluntly, dyers produce some of the most evil and obnoxious smells in our cities. And that's saying something, isn't it? Because let's be honest, with slop buckets emptied into the street and unrefrigerated meat and fish hung from hooks along narrow streets, our cities are not exactly sweet-smelling in the first place.

That powerful whiff, do you smell it? That's woad. You may well have smelt it before, because woad is the most common dyestuff in Europe. A pity then, that it's also one of the most evil-smelling. Imported as bales of leaves, dyers knead the woad into tight balls which they leave to ferment for nine weeks. Yes, you heard that right: nine long, smelly weeks. The stench is overpowering. They say the English Queen has banned woad processing anywhere near royal palaces – and by 'near', she means within five miles.

And that smell now ... Not much better, is it? It comes from thousands of tiny whelk-like sea creatures called Murex or Purpura molluscs that are slowly decomposing in that great big vat over there. When they are sufficiently putrid and decayed, they secrete a yellowish liquid, a bit like pus except that it stinks more. Yuk! But spread this horrible yellow stuff on woollen cloth and you get a rich royal purple colour.[1]

Ah, now this smell you'll be familiar with, of course – the stomach-churning smell of a particularly ghastly toilet. That's because those vats over there are full to the brim with human urine. Stale human urine. Gallons of the stuff. The dyers use it to remove iron salts from the alum before using it as a mordant to fix the colour in the cloth. The iron salts, you see, would otherwise leave dark stains in the cloth. The dyers use anything up to ten pounds of alum for an item of red or scarlet cloth, and need a correspondingly large quantity of human urine.

And there's another interesting mix of smells. Strident? Certainly. A bit fishy? And oily? You're quite right. It's actually a mix of smells, a *melange* of rancid oils, animal urine and brine of pickled fish that dyers use to vary the colour or improve the quality of the dye.

Oh, I wouldn't rub your eyes, if I were you – this stuff here is arsenic, and that tub over there is full of sublimated quicksilver, or mercury. Highly corrosive. In fact, it's probably best not to breathe in too deeply. The fumes you see rising here are from lye, which we obtain by using baking soda on ash. Lye, I should warn you, is strong enough to corrode metal. I shouldn't want to try it on skin.[2]

Alum produces acidic fumes, too.[3] And the dyers just have to put up with it. I mean, just listen to this recipe that we use to create yellow:

> *First take thy alum and pound it small. Put it in water and set it on the fire and stir. Then take thy cloth, wet it in another water and put it in the water that thy alum was melted in ... When thou alumest thy cloth [rub your cloth with alum], take a light linen clout [scrap] and alum it with thy cloth, for when thou makest thy colors thou mayest assay [try out] by thy clout if it be good or not.*[4]

Nothing there to suggest that alum workers should take any steps to protect their lungs or their skin, is there? I wouldn't like to know the damage that the fumes are doing to their nostrils and lungs ...

What, really? You'd rather end your visit? A pity, there are so many more fascinating products I could show you ...

But maybe I could interest you in a different alum-based trade? Look, there's a tannery over there. You can see the hides that have just been brought in – with little bits of flesh still attached to them. These will be scraped off before the hides are soaked in a bath of alum, salt, flour and egg yolk and hung out to dry. Then, before being tawed (i.e. dyed with alum), the skins are washed in a tub of stale urine. Well, all right, I suppose washed is maybe not quite the right word...[5]

Shield your eyes!

Out in the streets and squares of our northern Italian cities, your eyes will be dazzled by a riot of colour. For generations now our dyers in Venice, Florence, Milan, Lucca and other cities have had ready access to a dizzying variety of dyestuffs from the Mediterranean, Africa and Asia, and, just as importantly, to the highest quality alum from Phocaea, Koloneia and Egypt. As a result, each workshop has developed its own recipes of dyes, mordants and other ingredients to produce the most extravagantly luminous colours.[6]

Couple the dyers' skills with endless supplies of locally made cloth woven from the best English and Spanish merino wool, and top-quality cloth imported from Flanders, and you get two things. One is a large population: we are the most urbanised part of Europe. Venice, Milan, Florence and Genoa are four of the five largest cities in Europe, way ahead of Paris, Rome or London. And the second thing is vast wealth. Just look up at the buildings in our cities! And I don't just mean the churches and cathedrals: the Church is powerful enough to build fine cathedrals even in the poorer parts of Europe. No, let your eyes dwell on the magnificent civic buildings – the city council buildings, the libraries, the merchants' residences, the market buildings, even the shipbuilding yards. And, above all, feast your eyes on the colourful clothes you see all around you. Europe will never see the like again!

We get the deep translucent blues from indigo. It comes from India as a paste which our dyers dissolve in honey and lime to obtain a more brilliant, more beautiful blue than you'll ever get from woad. The bright orange is from brazilwood, from India, Ceylon and the islands we call the East Indies. And you see all those rich reds, more luxurious than can ever be squeezed out of madder? They're from kermes, produced from the dried bodies of a particular female scale insect. Like many dyes, kermes originally came from India, but now we breed the insects and manufacture the dye here in Italy.[7] And you may have noticed that 'sanguine' is in fashion at the moment. You get the colour by dyeing the cloth blue in woad and then soaking it in a bath of kermes. The result is the colour of dried blood – trust me, it looks better than it sounds.

Nor is colour the preserve of women. Both sexes like to show off as many colours as possible. If a man's cloak has one colour, its lining will have another. Rich women wear up to five layers of clothing and they want you to know it, so strategic slits in the front and sides of their dresses reveal the colour of the clothes worn underneath. Hats and girdles, shoes and ribbons all add more colour. In fact, with more and more middle-class people indulging in colour, flouting the laws that forbid them, the seriously rich are finding it ever harder to keep ahead. Their latest attempt is parti-coloured clothing – where the colour of the left-hand side of a man's or a woman's clothing is different from the colour on the right-hand side. From there you can go to quarters or eighths, or chequered clothes like a harlequin, with the super-rich dolling up their servants in their family colours, with the left-hand sleeve and right-hand legging one colour, the right-hand sleeve and left-hand legging another, and the body part a third colour.[8] Where, I wonder, will this all end?

Ah yes, you tell me the Medici, the richest bankers in Florence, and therefore in Europe, scorn this blaze of colour, and dress in black instead. That's true. But be not fooled, my friend: the Medici are no less showy than anyone else. For theirs is a rich and lustrous black, as far as you can imagine from the drab black sackcloth worn by friars and hermits. The Medicis' silks and velvets are woven with wool spun from dark-haired sheep, and only the Flemish know how to achieve the shiny jet-black from woad, madder and of course alum.[9] So, no: far from being a symbol of austerity, the Medici black is a public statement that they have the means to import the cloth for their clothes directly from Bruges. Braggarts!

Not that our weavers and dyers can't make hard-wearing, practical clothes too, you know. In fact, you may well have a much-loved item in *your* wardrobe which owes its existence to our Genoese dyers. For our Genoese weavers use a weave of wool and cotton to make a tough, relatively cheap and partially water-proof fabric, and our dyers then treat this fabric with alum and indigo to achieve a particular shade of pale blue. The result? *Bleu de Gênes* (Genoan blue), otherwise known as 'blue jeans'.

I bet you never realised that, thanks to alum, our seamen wore jeans back in the late Middle Ages, did you?

Block your ears!

For obvious reasons, our urban populations are only too aware of the awful conditions in which dyers and tanners carry out their trade – they can't not be. The smell is too strong. The Cairo Genizah, that treasure-trove of documents spanning a thousand years of Jewish life in Cairo, includes a court case brought against a dyer for the evil smell he causes as early as the 11th century.[10] So our dyers and tanners now reside in quasi-ghettos on the edge of towns, or along the river. That way, the noxious effluents from their businesses do not pollute the cities.

But if you're expecting a wave of sympathy for the workers who labour in such conditions, forget it. Associated as they are with foul smells and pungent fumes, dyers and tanners are looked down on by fellow artisans.

And unfairly treated. Want an example? In our day and age, it is essential for each craft to be represented by a guild. Guilds control their members' working conditions, their rates of pay and the price of their finished products, and they uphold the quality of their members' products – a weavers' guild, for example, might insist that only high-grade alum from Anatolia can be used, and not the lower grade alum from Spain or North Africa.[11] In a word, guilds fight for their members' interests in every way. But here's the thing: in an important cloth-making city like Florence, in which about one-third of the active population is employed in the cloth trade, there's a powerful Weavers' Guild and an equally powerful Wool Merchants' Guild (each rich enough to possess its own five storey (!) stone building)[12] – but a guild of dyers? No. Doesn't exist. What this means is that the weavers' and wool merchants' guilds keep the cost of dyeing as low as possible – at the expense of the dyers.

What's more, you'll hear all sorts of negative things about dyers in our cities. Rumours of immoral goings-on in the murky parts of town in which they live and work. Things like that.

But my advice is: shut your ears and ignore what you hear. For dyers are as skilled as any other artisan. Each dyer has years of training behind him. The apprenticeship alone can be three years for plain dyers, anything up to seven years for dyers of more complex colours, and at the end, there's an exam at which the apprentices have to produce samples of their work. If they pass, they spend three years travelling from town to town, working for different masters and picking up the tricks of the trade.[13] Only then can they even think of applying to become a master themselves. So, given that our average life expectancy (in the 1400s) is 40 years or less, you can see dyers have spent a hefty portion of their lives in training. They deserve better than their reputation.

All the above was true in the first half of the 1400s.

But by the late 1450s, all this was at risk. With the cost of alum from Turkey and Egypt spiralling and supplies dwindling, the Genoese alum trade was close to collapse. And if it went, Europe's cloth industry, with its wealth, its expertise and its colour, would come tumbling down with it. No alum, no colour. And no colour was unthinkable.

Alum was of course not the only oriental commodity in short supply. There was a shortage of grain, salt, spices and exotic fruits, but most of these could be sourced from other countries. Not so with alum. Everyone tried, of course. The Genoese re-opened the alum mines in Volterra and Piombino in Tuscany and brought alum from the tiny islands in the Gulf of Naples. The Venetians went looking for alum in Tyrol.[14] Alum from Sicily was traded in Genoa for the first time in 1458. The Catalans shipped alum from Spain and from Tripoli in North Africa. But it was all low-grade stuff. In no way could it compare in quality or quantity with the alum from the eastern Mediterranean.

Threatened with bankruptcy and unemployment, in October 1458 the dyers of Genoa sent a delegation to the city authorities to plead for increased imports of alum. Their work, said the dyers, was becoming impossible. And the authorities were so concerned that they authorised a desperate measure, and decreed that every Genoese ship with alum in its cargo should call at Genoa and off-load 50 cantara of alum, even if it had been ordered for delivery elsewhere.[15]

Of course, this measure only served to undermine any remaining trust that customers might have in Genoese merchants, and exacerbated the shortage of alum in all other cities. Nor was it even remotely enough to solve the crisis in Genoa. Ships that had plied the Mediterranean lay idle. Families who had staked their wealth on the alum trade were staring ruin in the face.[16]

Imagine the fury then, when Genoa heard in this same year of 1458 that three English ships led by a Bristol merchant called Robert Sturmy, and financed by a group of wealthy English investors, had for the first time ever entered the Mediterranean with the intention of trading in what had traditionally been Italian stomping grounds. The English ships had off-loaded good Cotswold wool in Pisa and were at that very moment loading up in Chios with alum for the English market. One hundred and eighty years after the Zaccarias had first shipped alum directly from Phocaea to Southampton, the English were now seizing this trade for themselves. And in a big way. Robert Sturmy's three vessels took on a colossal 152 tons of alum, enough to supply the needs of the English cloth industry for one whole year. If Sturmy's expedition succeeded, it would be a game changer for English trade.

But for the Genoese, this was more than just a snub. At a time of dwindling stocks, it looked like the final nail in the coffin of Genoese dominance of the alum trade. It was not to be countenanced. So an ambush was organised, and as Sturmy's ships skirted Malta on their homeward journey, they were set upon by 'pirates', led by a Gattilusio, whose family ran the alum mines on Lesbos. And the attackers were in no mood to show mercy: 128 English sailors and merchants were killed, including Robert Sturmy himself, and two of the three English ships were sunk.

The Genoese had vented their revenge, and the blow to the English investors and indeed the English economy as a whole was massive, equating in today's prices to a loss of £750 million. What's more, the English merchants now demanded revenge of their own, and persuaded King Henry VI to have the entire (and entirely innocent) Genoese community in London thrown into prison.[17] The high cost and diminishing reserves of alum had driven two normally friendly trading nations to the brink of war.

All over Europe, everyone was praying for a miraculous discovery of a new source of high-quality alum, closer to home. To keep the Ottomans at bay. To prevent trade wars in Europe. And simply to make it possible to carry on wearing all those lovely colourful clothes.

Could the miracle happen?

10

Alum, the Pope and a Scam

The miracle happened.

In 1460 a new source of alum was discovered. And everything about it was miraculous.

The new source was not only located in western Europe, but in Italy. Not only in Italy but in the Papal State, the large area of central Italy that was directly governed by the pope. And not only in the Papal State, but in Tolfa, only 60 kilometres from the very gates of Rome. It was top quality alum, too, and in amounts that were beyond big: seventeen *million tons* of alum would be mined at Tolfa between 1462 and 1796.[1] That would put even Phocaea in the shade. And discovered within a decade of the fall of Constantinople: what could be more miraculous than that? To cap it all, the man who stumbled on the alum deposits, Giovanni di Castro, was related to the new pope, Pius II. Or was maybe his godson. It really didn't matter. Everyone got the central point. Alum, once the money-spinner for Orthodox Constantinople, more recently for the Muslim Ottomans, had now fallen into the lap of Rome.

Before fleeing Constantinople in 1453 Castro had run a dyeing business there so he was familiar with alum. And the story goes that while prospecting in the hills northwest of Rome in 1459 or 1460, he happened to notice a type of grass which he knew grew near the alum mines in Turkey. Intrigued, he picked up some of the local rock, licked it, found it to be salty, and ran to the pope to tell him it must be alunite.

We too may take some of the details of this story with a little pinch of salt. Pope Pius himself was sceptical and had samples of the rock tested by experienced dyers in Genoa and Venice. But, to his surprise and delight, the dyers confirmed that the rock was indeed alunite, with an extraordinarily high alum content. There was universal jubilation in

Rome, people fell on their knees and thanked God for this heaven-sent boon. The dearth of alum was resolved. The reliance on Muslim alum was over. Over the next three centuries, alum from Tolfa would fill the holds of ten thousand ships and a never-ending stream of revenues would fill the coffers in Rome.[2]

Production swung into action, immediately and on a grand scale. The Frangipani family who held the land on which the alum was found were paid off.[3] Housing was put up for the workers, the port of Civitavecchia was extended to handle the alum exports, new roads were laid out from the harbour to the alum quarries. A new chapel would see to the miners' spiritual needs. Wood from the extensive Tolfa forests was reserved for burning the ore: once production was in full swing, the alum mines would require 30 000 cubic metres of wood a year to keep the fires burning.[4] This massive project was managed by the Apostolic Chamber, the papal office that collected income from rents, bequests, sales of indulgences, monastic production, etc. The Chamber, which dealt with more money each year than any government in Europe save the Ottomans in Constantinople, was experienced in dealing with high revenues.[5] A cargo of Tolfa alum bound for Flanders and England left Civitavecchia as early as 1463.[6]

From the very beginning, the production of alum at Tolfa was publicly hailed as a crusading, anti-Ottoman project. 'Today I bring you victory over the Turk,' Giovanni di Castro told Pope Pius II when he first discovered the Tolfa alunite. 'If you will give orders to engage workmen, build furnaces and smelt the ore, you will provide all Europe with alum and the Turk will lose all his profits ... Now you may equip a war against the Turks.'[7]

It was a two-pronged strategy. Part one was to inflict a painful financial loss on the Ottomans by ending the trade in Turkish alum. Thus in 1463, casually mixing temporal interest with spiritual authority, Pius II banned Christian merchants from importing Turkish alum under pain of ex-communication. Anyone found buying Turkish alum would be banned from church, could not be married, and would receive no final rites when they died, meaning that their soul would go to hell. Tolfa alum was as important as that.

Part two of the strategy was to keep the income from the sales of Tolfa alum in a separate Vatican account, earmarked to fund 'a great crusade against the Sultan of Constantinople'.[8] Indeed when Pope Pius II died in 1464, the cardinals who assembled in Rome to pick his successor used their first day to confirm that all monies from Tolfa must be used to wage war against the Ottomans, postponing the election of the new pope to the second day.[9] Not that they need have worried, for in his Maundy Thursday address in 1465 the new Pope Paul II not only confirmed his predecessor's ban on importing Turkish alum, but widened the scope of the ban to apply to the trading of *any* alum from *any* source other than Tolfa, including alum from North Africa, Italy, Greece and Spain.[10] This was justified on the grounds that the proceeds of sales of Tolfa alum would be spent on funding a crusade which would be in the interests of all Christian Europe.

What's more, in 1466 Pope Paul II brought in new management know-how in the form of the richest bankers in Europe, the Medici of Florence. The Medici were to oversee the production of Tolfa alum, set its price, and guarantee the Pope a cut of two ducats on every cantaro of alum that left Civitavecchia.[11] Their appointment made sound commercial sense, since the Medici had agents in Flanders and in England, the two major alum markets. But it was also one in the eye for the Genoese who, with their experience in the alum trade, might have been expected to pick up the contract. But this was payback time. The Genoese had long flouted the Pope's orders to cease dealings with the Muslim 'enemy': in Chios, in fact, they were still paying the Turks 12,000 ducats a year.[12] Now, thanks to Tolfa, the Pope could call the shots, and the contract with the Medici was extended for a further four years in 1471.[13]

Was there even possibly a third strand to the anti-Ottoman strategy, namely the putting of Muslim slaves to work in the papal alum mines? Quite probably. We know that Muslim slaves were used in public works in Rome, and were exploited in the Tolfa alum mines at a later date. But I have not been able to ascertain when the practice began.[14]

With the world's largest known alum reserves firmly in papal hands, the running of the mines controlled by Europe's most powerful banking family, and a monopoly on all sales in Europe, Rome's crusading fund was set to soar. There was hope that the Christian west might yet to recoup the lands that Christendom had lost in the east.

◆◆◆◆◆

And indeed production leapt, sales of Tolfa alum flourished, and money poured into the Pope's coffers. But there was, mercifully, no crusade.

Monarchs in Europe still regularly expressed their Catholic credentials by mouthing support for wars against the Ottomans, but they had largely lost their appetite for these pointless military campaigns. And so had their peoples. There was less blind faith in the church and more prosperity that people were loath to squander. What's more, for all their belligerent talk of crusades, the popes themselves found it just too tempting to spend their sudden, unexpected and wonderfully massive fortune on pet projects of a more peaceful nature. Pope Pius II rebuilt his home town of Corsignano in a modern Renaissance style, even changing its name to Pienza, in memory of himself. Paul II, forgetting his promise to keep the income from Tolfa in a separate war account, used it to indulge his love of art and antiquities. His successor Sixtus IV spent 100,000 ducats of it to buy himself a jewel-studded tiara.[15]

As a result, while a number of local fights against the Ottomans were hyped by the pope as the dawn of a new Christian onslaught on the infidels, in truth they amounted to no more than local powers – the Venetians, the Wallachians under Prince Vlad, or the Albanians under their leader Skanderbeg – settling local scores with the Ottomans, and happily taking the credit if the pope wished to call their actions a crusade.

However, crusade or no crusade, the popes maintained their claim to a monopoly on all sales of alum in Europe, choosing to ignore the fact that canon law took a dim view of monopolies.[16] And this led to clashes with the other producers of alum who were by no means willing to keel over and sacrifice their lucrative income just so that the pope could increase his. There were alum mines in Castile and Aragon in Spain, in Tuscany and on islands in the Bay of Naples,[17] Venetians ran alum mines in Crete and, for a while, in Phocaea,[18] and Catalan, Genoese and Venetian merchants traded alum from North Africa. One and all, they made it clear that they would ignore the Pope's decree. The Medici and the Vatican were therefore forced to face the fact that the mere declaring of a monopoly would not work. They would have to resort to other methods of persuasion, taking local circumstances into account.

An alum producer in a strategically weak position could be eliminated by force. This explains why in 1472, in an act of unprovoked savagery, Medici-run Florence launched a vicious attack on the semi-autonomous Tuscan city of Volterra. Its troops besieged and sacked the city, raped or slaughtered many of its hapless inhabitants, all in order to close Volterra's alum mine. The brutality was out of all proportion to Volterra's very modest output of alum, which was no real threat to Tolfa at all,[19] but the Pope and the Medici wanted to show they were serious about imposing their alum monopoly, even to the point of spilling blood. Yes, Tolfa alum was as important as that.

The alum mines in Agnano and on the island of Ischia, both in the territory of Naples, were a very different kettle of fish. Naples was a dependency of the King of Aragon, a powerful and potentially dangerous rival to the Papal State in Italy so warfare was out of the question. Instead the pope suggested a mutually beneficial business deal: why not split the alum market in Europe between them? Rome and Naples could agree joint production targets, sell their alum at an agreed price, and allow on-site inspectors in each other's alum mines to see that the deal was being adhered to. Both states would attack any ship carrying alum that was *not* from Tolfa or Naples, and as prisoners taken from such ships would be deemed heretics for defying the papal ban, they would be killed.[20] John II of Aragon, needing funds to fight off a rebellion back home, signed up in 1470, with the treaty intended to last 25 years.

But this classic cartel arrangement, designed to secure a high commodity price by rigging the commodity market, lasted nowhere near 25 years. The Medici soon discovered that Naples produced such little alum, and its quality was so poor, that it constituted no threat to Tolfa, and Pope Paul tore up the treaty a little more than a year after it came into operation.[21] It was, moreover, sweet revenge on Ferdinand, the cocky heir to Aragon who, when he had married Isabella of Castille the year before, had not only failed to obtain the necessary papal dispensation needed for marrying his second cousin, but had infuriated the pope by fabricating a false one.

The pope, however, was focussing on the wrong side of the royal Spanish wedding. The real rivals to Tolfa were not the Aragonese mines in Naples but the much larger alum reserves at Mazarron in Castile, discovered in 1462 (the same year as Tolfa) and destined to become the second largest producer in Europe.

Quite apart from the issue of rival alum production, the Vatican faced the problem of illegal sales from Ottoman Turkey, and here the pope and the Medici were victims of their own success. The aim of the monopoly had been to drive up the price of alum, and by 1466 the price of Tolfa alum had indeed risen to 3.25 ducats per cantara. But the higher the price of Tolfa alum, the greater was the incentive for the traders and dyers to buy the cheaper Ottoman alum, whatever the peril to their souls. So alum smuggling boomed, and large quantities of 'illegal' alum from Trebizond, Chios and Phocaea found their way to Bruges and other outlets.

Lacking a fleet that might police the sea routes, Rome tried offering European monarchs financial incentives to quash these 'illegal' sales within their realms. In 1467, for example, Rome offered King Edward IV of England a share of the profits from the sale of Tolfa alum in his kingdom if he agreed to stamp out the trade in cut-price Turkish alum. Mid-way through the Wars of the Roses, Edward would dearly have loved the extra income; but he ruefully turned down the offer for fear of antagonising his English wool and cloth merchants who would, of course, face higher prices for their alum if he cut cheaper imports.[22]

Rome had more luck in Flanders, however, where the pope had a degree of leverage over the ruler, Duke Charles the Bold of Burgundy, who needed a dispensation from the pope to marry his fourth-degree cousin Margaret of York, the sister of King Edward IV. Usefully, too, Charles was in debt to the Medici for huge loans supplied by Tommaso Portinari, their agent in Bruges. So when Portinari put it to Charles in 1468 that the dispensation might be procured and the debt overlooked if he agreed to ban the sale of non-Tolfa alum in Burgundy and in Flanders, Charles signed up. The deal was to apply for a period of twelve years.

Twelve years? Some hope! There was uproar and outrage in Flanders when the contents of the deal became public, with the dyers and cloth merchants in Bruges, Ypres and Ghent taking to the streets, ridiculing Rome's assurances that the price of Tolfa alum would be kept below the price of alum in neighbouring countries. The alum merchants joined in when it transpired that the deal banned them from selling off any non-Tolfa alum that was already in their possession. Of course, in practice, they simply sold off their existing stock of alum illicitly. Indeed Tommaso Portinari himself, never one to miss a trick, bought up some of this 'illegal' stock, re-branded it as Tolfa alum, and then re-sold it at the higher Tolfa price.

The anger in the streets was so real and the mutterings of rebellion so loud that Charles delayed implementing the treaty for eighteen months, and when in 1473 the States General (Parliament) reduced his dependence on the Medici by agreeing to cover the costs of his army for six years, the Duke rescinded his alum deal with Rome and allowed the import of all 'Christian' alum once more. Charles had played his cards astutely. He had got permission to marry his Margaret in 1468, and his Flemish subjects had coughed up the money he needed for his army. It was not only the pope and the Medici who could play the alum card to their own advantage.[23]

Rome was having to learn that an absolute monopoly on alum sales was impossible to achieve, and that a degree of flexibility could be more effective. When negotiations with Venice collapsed in 1469, for example, because the price of three ducats per cantara was more than the dyers in Venice could swallow, the Vatican came back with a new offer of 2.2 ducats per cantara and a deal was struck. What's more, as the deal only applied to sales by land, the Vatican was tacitly turning a blind eye to imports from Crete and Turkey. Rome had learned that it could make money by selling alum to Venice while keeping 'illegal' imports at a level that it could live with.[24]

And this was the delicate balancing act that Rome eventually settled for. The reserves of Tolfa alum were so enormous and of such high quality that they were bound to dominate the European market, monopoly or no monopoly. Rome could therefore put up with sales of alum from, for example, Mazarron in Castile, and with illegal imports from Turkey and Egypt, as long as they were not excessive.

Thus, in the year 1480, we find a Venetian ship loading alum in Phocaea, blithely ignoring Pope Sixtus IV's ban on trading non-Tolfa alum, despite his threat to consider traders in such alum as heretics and excommunicated from the Church. The Genoese, too, continued to import Turkish alum.[25] And such relatively small-scale smuggling hardly affected the high price of alum. The price might fall locally if two shipments happened to arrive at the same time, causing a sudden glut, or might rise if a cargo was lost at sea due to bad weather or an act of piracy. A galley with 820 cantara of alum on board was lost in March 1464, for instance, a second ship was shipwrecked off Boulogne in October of the same year and in 1465 a shipment of Tolfa alum expected in Flanders in the spring did not arrive till the autumn as a result of stormy weather.[26] Taken in the round, however, the price of Tolfa alum remained stable.

Till 1486. For in that year, an alum smuggling racket came to light that was on a much more massive scale than anything attempted before. It was an operation clearly sanctioned from the very top, and at its centre was King Henry VII of England.

It all began with the curious case of the kidnap of a Spanish boat. In 1486 Giovanni da Negroni, a Genoese living in England learned that a cargo of alum from the Tuscan port of Piombino (i.e. not from Tolfa) was on its way to Flanders, and paid a band of thugs to board the ship and tow it to England. Then, when it arrived, Negroni laid claim to the alum on board, on the grounds that it was 'illegal' according to Sixtus IV's ban on sales of non-Tolfa alum. It should, he said, be taken straight to Rome. With a cut for Negroni's good Christian self, of course.

No! Wrong! countered an unnamed Florentine merchant, who strangely and suddenly appeared before the king. Peculiarly well-versed in English law, this Italian merchant protested that whatever the pope had said, there was no law in England against the alum trade. On the other hand, given that the vessel in question was from Spain, with whom England was in alliance, and that some of the Spanish crew had been killed, it followed that Negroni was guilty of causing the deaths of allies of the English crown and should be made to pay for the crime.[27]

Nonsense! retorted Negroni. Under the terms of Pope Sixtus's edict, traders in non-Tolfa alum were heretics. By supporting the heretic traders, the Florentine merchant had himself become a heretic, and the opinion of a heretic was void in a court of law.

King Henry sagely put the case into the hands of his justices. They considered the case, and when they gave their verdict, it went against Negroni. This enraged Pope Innocent VIII, who reminded Henry that when he had seized the throne only the year before, he had promised the pope his eternal loyalty. That loyalty, said the pope, demanded that he intervene in the case. To which Henry responded with rare Tudor humility. He, Henry, had of course the greatest sympathy for the pope, but he could not, he explained, as king, intervene in a court of law, all the more so because – as the pope himself had pointed out – he, Henry, was 'new to the kingdom'.[28]

There was of course more here than met the eye. The mysterious unnamed Florentine 'merchant' who countered Negroni was in all likelihood Ludovico della Fava, agent of the Florentine Frescobaldi Bank and Henry's personal banker. Or someone paid by della Fava. And Negroni himself had probably been set up to kidnap the Spanish ship in the first place, because this was no single shipment but part of a wider racket. Nor was the court as independent as Henry maintained. Henry was clearly allowing non-Tolfa alum to come into the country at a low price in order to advantage the English textile industry, and the Negroni case did not put an end to it because a Venetian alum ship is recorded leaving Chios for Flanders and England in April 1487.[29] Indeed Henry himself had his nose in the trough because he was charging customs duty on the 'illegal' imports.

And this was only the beginning. The smuggling racket was ramped up when Julius II became pope in 1503 and hiked the price of alum. It soared 60% in England between January and December 1504, and in the Netherlands, it rose from 42 shillings per load (= 188 kg) in 1501 to 120 shillings in 1504.[30] The result was an immediate, painful economic recession with violent protests in the streets of Bruges and Antwerp, and the rulers – now the Habsburgs, who had taken over from the Burgundians – again fearing a rebellion. All of which Della Fava saw as a wonderful business opportunity, and he approached Henry with a new idea ...

Della Fava's plan – now known as the 'Frescobaldi deal', after the Italian bank that financed the racket – was simple enough. The Habsburg rulers of Flanders feared a revolt triggered by the exorbitant increase in the price of alum, but could not undermine the pope and import Turkish alum directly, as they wanted papal support for having a Habsburg crowned Holy Roman Emperor. They would, however, according to della Fava's agents in Flanders, turn a blind eye to any Turkish alum that might be smuggled into England and re-exported to Flanders as 'Christian' alum. If Henry increased his imports of Turkish alum, therefore, the Flemish dyers would have their cheaper alum, the Habsburg rulers would be in Henry's debt for defusing a potential uprising and Henry could make fabulous amounts of money by charging duties on the imports, re-selling the alum at a higher price than he had bought it for, and then again by charging duties on the exports to Flanders.

The plan would of course require larger cargoes of alum, so Henry's involvement in the smuggling racket would increase from simply allowing it, to providing the ships that made it possible. But, Della Fava assured the king, his role would not be made public. He could simply lease two English ships to the Frescobaldi company in Florence, who would fly their company flag when using the ships to transport the alum. Henry could therefore protest his complete innocence and deny all knowledge of the operation. Oh, and he could make even more money by charging for the lease of the boats.[31]

Henry VII was one of the very few kings of the age who did his own accounts and one of the few who understood that money gave a ruler freedom of action (impoverished monarchs had to call parliaments to vote them funds, and parliaments then exacted concessions). So of course Henry acquiesced, and willingly. Two large ships, the 1,000-ton *Regent* and the 800-ton *Sovereign*, registered as belonging to the Frescobaldi Bank, sailed to Chios (the first English ships to do so since Robert Sturmy's fateful expedition) and returned bulging with Turkish alum – some of it for dyers in England, but much for re-export to Flanders, and to France and Spain as well. The smuggling racket worked to perfection, and the price of alum fell from 120 to between 40 and 60 shillings per load.[32] There was satisfaction all round.

Except, of course, in Rome, where the profits made by Tolfa alum fell accordingly. The amounts involved were now too large to be overlooked, so in 1506 Julius II renewed his ban on all imports of non-Tolfa alum. He also used cutting-edge technology to have it published as widely as possible: the ban on sales of alum was the first European law printed on a printing press. And the ban showed Julius knew perfectly well who was behind the smuggling ring because it named Nicolas Waring, a crew member of the *Sovereign*, and Ludovico della Fava himself.

To no avail. The smugglers continued as before. In December 1506, therefore, Julius ordered Pietro Griffo, his nuncio, or ambassador, in England, to complain directly to Henry. Henry listened with due decorum. But without undue alarm, for he knew (and as Julius did not) that Griffo was himself in on the alum racket and taking a cut.[33] But Henry again wrote to the pope protesting his innocence and assuring the pope of his loyalty in the time-honoured way: he urged the pope to launch a new crusade and even offered (tongue in cheek, surely?) to provide men and money for it.

All the time, of course, Henry was wily enough to continue importing decent amounts of Tolfa alum to keep up the pretence that England depended on it. He even fined any merchant (apart from his own, of course!) caught importing non-Tolfa alum into England. Henry could therefore demonstrate that he was taking action to protect the sales of Tolfa alum, while pocketing both the fines and the profits on his own illegal imports.[34]

The sums involved in the scam were colossal. Customs duties of thirteen shillings and fourpence were charged on each *cantara* of alum, so a shipment of 13,000 *cantara* made Henry a cool £8,666 in customs dues alone. Henry's takings of over £15,000 on just one transaction in the scam was the biggest single mercantile deal in his reign.[35] Such staggering figures show that the exports of alum from England to the Low Countries must have been significant, reversing a history of alum being exported the other way.[36] It has indeed been claimed that the proceeds from the alum racket were England's 'chief source of revenue outside of taxes'.[37]

The wider impact of the Frescobaldi alum deal was significant, too. In Flanders, it defused the political crisis and kept the Habsburgs safely in the saddle. In England the cloth industry benefitted from its access to cheaper Turkish alum and exports of English cloth leapt 61% during Henry's reign.[38] England had been broke at the end of the Wars of the Roses; the Frescobaldi alum deal put its economy back on a sound footing.

And in so doing, the alum racket also strengthened Henry VII's personal grip on power. He only called seven parliaments during his entire reign and was therefore largely free to act as he wished. Henry's alum-fed treasure chest gave him the means with which to see off rival contenders to his throne (Lambert Simnel, Perkin Warbeck and others) by bribing monarchs in Europe to stop aiding them. A sweetener of £10,000 in 1502 persuaded the Habsburg Emperor Maximilian to stop supporting any English rebels; the incredible sum of £108,000, equivalent to Henry's ordinary annual income, brought a similar undertaking from King Louis XII of France in 1505;[39] and a 'loan' of £138,000 secured a promise from Philip of Burgundy, Maximilian's successor, to hand over Edmund de la Pole, yet another claimant to Henry's throne.[40] Payments on this vast scale would not have been possible without the stream of income made from the alum sales.

The Frescobaldi alum scheme therefore helped to secure the Tudor dynasty's hold on England. By the time Cardinal Wolsey put an end to it, Henry VII's son Henry VIII was in firm control of the country.

And alum also transformed the Vatican and the papacy.

Prior to 1462, the papacy had ever been on the brink of insolvency, with popes unable to assert themselves on the European stage because they were financially dependent on the goodwill of this or that European monarch. Tolfa alum changed all that. The popes were now able to grace their city with ornate buildings and monuments, and their palaces with paintings and sculptures. Their lavish ceremonials could now compare with those of any court in Europe, and they could pursue a more active, and more aggressive, policy within Italy.

Just consider the figures. In 1473, ten years after the first shipments of alum from Tolfa, alum accounted for 100,000 florins out of a total annual income of 300,000, so fully one-third of the Pope's income derived from the sale of Tolfa alum. And, in addition, Pope Sixtus IV was able to borrow a further 62,918 florins from the Medici bank on the surety of the Tolfa alum mines. 'The value of the alum monopoly to the papacy,' it has been rightly said, '... cannot be overestimated'.[41]

And that was only the start. As alum production soared from a little over 900 tons a year in 1466-1479, to nearly 1,300 tons by the end of the 1400s, and nearly 1,800 tons a year in the mid-1500s, the pope's income rose correspondingly.[42] In the 1470s and 1480s, Sixtus IV had an annual income of 290,000 ducats a year; some forty years later Leo X was enjoying an annual income of 420,000 ducats, a 45% increase. Another source estimates that papal income rose three or fourfold between 1480 and 1599.[43]

Of course, alum was not the sole source of papal income – the popes also had a monopoly on sales of salt, for example, and they sold licences for the export of grain. Nevertheless, the impact of alum on papal income was massive, and gave the popes of the 15th and early 16th centuries the opportunity to get their finances under control.

Which, however, they did not do.

It was, of course, just too tempting to spend this wonderful new income which had appeared so unexpectedly. So as the popes' income grew, so did their expenditure. The popes of this era transformed the Vatican into a Renaissance palace and lived the life of Renaissance princes. They acquired fabulous art collections and commissioned Raphael, Bramante, Michelangelo and other top artists of the age to design and decorate the chapels and palaces that we see and admire today. It was in this period that the Sistine Chapel was painted, that St Peter's Cathedral was built and, more darkly, that the popes went to war. The Vatican's two biggest items of expenditure, Charles Stinger writes in his book 'The Renaissance in Rome', were 'papal wars and pontifical splendour'.[44] The popes financed armies and bribed allied powers in the wars that raged in Italy for much of the 15th century. Sometimes they even took part in military action themselves, as when Julius II personally took charge of the army laying siege to Mirandola in 1511, in an effort to keep French forces out of the Po valley.

Only one other person in Rome could match the popes in terms of extravagant spending, and that was the new manager of the Tolfa alum mines, the Sienese business tycoon Agostino Chigi. He had secured a twelve-year contract in 1500, having of course paid the pope handsomely for the privilege,[45] and the revenues from Tolfa made Chigi the richest private individual in Rome, indeed one of the wealthiest men in Europe, with an estimated income of 70,000 ducats a year.[46] On the back of his income from Tolfa, Chigi built both a banking empire and the most luxurious Renaissance villa in Rome, where he flung sumptuous banquets, any one of which might set him back 2000 ducats. When Agostino Chigi died in 1520, the alum magnate reputedly had 20,000 men in his pay, a hundred vessels under his flag, and an estate worth 800,000 ducats. The chapel he was buried in had been personally designed and decorated by Raphael.[47]

Chigi, however, for all his lavish spending, was a savvy businessman with a keen eye on the bottom line. He knew exactly how much he could spend without getting into debt – which the popes, significantly, did not. They had little knowledge of, and even less interest in, finance, had only a hazy idea of the Vatican's bank balance and utterly failed to appreciate the increasing importance of money.

The income from Tolfa, therefore, which began as a miracle windfall that might have saved the Vatican debt, ended up feeding the popes' appetite for ever greater expenditure, with fateful consequences.

⁂

Quite apart from the sheer amount of it, the money made from sales of Tolfa alum was important because it was 'clean' money, untainted by corruption. This was unfortunately not necessarily the case with all sources of papal income.

One source of funds, for example, went by the name of 'sales of offices', and consisted of the substantial payments made by wealthy clerics in return for being appointed to senior offices in Rome. Bishops, too, paid generously for being raised to the status of cardinal. Such practices were of course open to abuse by a pope in need of money, and explains why Leo X appointed no fewer than 31 cardinals during just eight years in office.

Even more iffy was the selling of *indulgences*. The faithful had always paid to have prayers said in their name, but this practice went further: people paid for the promise that their sins were forgiven, and that their time in hell or purgatory would be shortened. This morally dubious practice grew so fast within Italy in the early 1400s that it might well have undermined the pope's moral authority,[48] but from 1462 the income from Tolfa alum removed the need to rely on indulgences, and the threat receded.[49]

Forty years later, however, papal expenditure had once again spun out of control. Pope Leo X's income of 420,000 ducats a year, massive though it was compared with that of his predecessors, was hopelessly outweighed by his outgoings of 600,000 ducats a year. By 1515 Leo's payments to 88 professors in Rome were in arrears,[50] but still Leo continued to live his opulent lifestyle. And then, in the fateful year 1517, Leo embarked on a ruinous war against Urbino. The cost of that war alone has been estimated by a contemporary historian, Francesco Guicciardini, at 800,000 ducats.[51]

Not even the spectacular sales of Tolfa alum could plug gaps created by profligacy on such a scale, and Leo was forced to resort once again to selling indulgences and offices.[52] Indeed, far greater sums of money now had to be raised in this way than before: Pope Sixtus IV, with his lower income, had made do with 40,000 ducats a year from selling indulgences and offices, but in 1517 Pope Leo X had to raise 144,000 ducats in this way.[53] Such massive sums could not be found within Italy, so Leo was forced to authorise the sale of indulgences in the German lands. The consequences, as we know, were a catastrophic loss of moral and spiritual authority in northern Europe which led directly to the Reformation.[54]

The role of alum had therefore been both benign and corrosive. On the one hand, the huge sales of alum had enabled the Catholic Church to rule supreme for longer, by deferring the need for widespread use of the corrupt practices and avoiding the moral outrage they might have sparked. On the other hand, the massive income from alum tempted the popes into a lavish lifestyle, gross expenditure and unnecessary warfare, thereby ultimately undermining the authority of the Catholic Church by exposing the popes as morally bankrupt.

⬩⬩❖⬩⬩

In 1517, the year that Leo launched his war against Urbino, Martin Luther wrote his *Ninety-Five Theses*, thereby effectively launching the Reformation. What had provoked Luther into action was his horror at hearing the Dominican monk Johannes Tetzel selling indulgences in exchange for a promise of divine pardon for all 'sins, transgressions, and excesses, how enormous soever these may be'.[55] For the message Tetzel was giving was very clearly, as Ginny Justice has put it, 'that one could do almost anything, and an indulgence could replace the penance for the sin'.[56]

Almost anything? A tiny number of sins were considered so terrible, so heinous that no indulgence could wash them away, no matter how much money was paid. And one among this small number of unpardonable sins was the sin of trading non-Tolfa alum. No indulgence could be granted for it, not even by Tetzel.[57]

Tolfa alum was as important as that.

11

Alum and a Refugee Magnate

When in September 1566 Selim II succeeded his father Suleiman the Magnificent as Sultan of the Ottoman Empire, one of his first acts was to reward those who had supported him by offering them the top posts in the administration. There was nothing surprising about this. It was standard practice, although it is true Selim maybe had more need of support than most, given that his path to the throne had involved the deaths of two of his brothers.

Nor will we be surprised, knowing something of the history of the alum mines in Turkey, that among the lucrative posts the new Sultan granted was the right to farm the alum mines of Foça, as the Ottomans called Phocaea.

What maybe comes as a surprise, however, is that the co-lessees of the alum mines of Foça were both Jewish. One of them, by the name of Yosef Nasi, was one of Sultan Selim's very closest friends and advisors. Yet strangely, Yosef had been brought up not in Turkey but in Spain, and under the name not of Yosef Nasi but of Juan Miques.

What was going on here?

⁂

This particular story begins in 1478, when the Catholic Monarchs Ferdinand and Isabella of Aragon and Castile viciously turned on a section of their own subjects, inviting the Inquisition to Spain to persecute them.[1] No matter that these families had considered Spain to be their home for generations, tens of thousands of Spaniards were – simply on the grounds of their religious beliefs – subjected to interrogation, imprisonment, torture and execution.

This evil policy culminated in 1492 in the expulsion of Spain's Jewish citizens unless they renounced their faith.[2] As a result, thousands of Jews left the country in a massive self-inflicted brain drain for Spain, while the thousands who stayed adopted 'Christian' names, outwardly led a 'Christian' life and practised their Jewish faith in secret. These people, who were called New Christians or, pejoratively, *marranos* (one meaning of which is 'swine'), lived in peril of being discovered by, or betrayed to, the Inquisition.

Yosef Nasi was born into one such family in 1524. His father, who had taken the family name 'Micas' (also spelt 'Miques' or 'Miguez'), died when Yosef was only one year old, so the boy was brought up in Spain by his uncle 'Francesco' and aunt 'Beatrice' alongside their daughter 'Reyna', and by his uncle 'Diego' and aunt 'Brianda'. The families had taken the name 'Mendes' (also spelt 'Mendez'). Ironically, even Joseph, the name of the man Jesus regarded as his father, was considered too un-Christian a name, so the little boy was called Juan in Spain, João in Portugal.[3] Remembering who was who in this bewildering array of names was crucial to outwitting the Inquisition.

'Francesco' and 'Diego' were seriously rich bankers, who lent money to some of the wealthiest families in Europe.[4] However, even this high status did not guarantee them protection from the Inquisition, so the two families moved from Spain to Portugal, then from Portugal to Antwerp, in Spanish Flanders, in search of safety. In each move they used their international banking contacts to make sure their wealth travelled with them.

'Francesco' and 'Diego' died in Antwerp, so 'Beatrice' managed the banking business till the Inquisition arrived there too. 'Beatrice', her daughter 'Reyna' and her sister 'Brianda' then moved to Venice, leaving 'Juan', aged 21, to run the business in Antwerp. 'Juan' organised the safe transfer of the family's wealth to Venice, no easy matter, seeing as 'Beatrice', 'Reyna' and 'Brianda' had broken the travel ban imposed on New Christians, and Emperor Charles V of Spain had his eye on confiscating the very attractive 'Mendes' family fortune. But 'Juan' managed to get the money out, in part by lending some of it, the massive sum of 150,000 ducats, to Henry II, King of France,[5] the understanding being that the repayments on the loan would be made to the 'Mendes' bank in Venice.[6] 'Juan' then followed his aunts and cousin there.

However, Venice was soon as unsafe as Lisbon or Antwerp, and in 1549 'Beatrice' was imprisoned and officially declared to be a Jew. This of course put the whole family in danger, so 'Juan', 'Reyna' and 'Brianda' fled to Ferrara, while King Henry of France stopped repaying his loan, under the convenient pretext that he had not known he was borrowing from a Jewish bank.[7]

Having run out of road in Christian Europe, 'Juan' put out feelers about moving the family to Istanbul. And the response was very different. For while Jews did not enjoy the same rights as the Muslim population, they enjoyed considerably greater safety here than anywhere in Christian Europe.[8] Indeed, they thought of the Ottoman Empire, as Marianna Birnbaum puts it, 'as East European Jews thought of America in the nineteenth: a haven from persecution',[9] and as a result Jews accounted for 40,000 or 50,000 of Istanbul's population of 250,000 in the 1550s.[10] Moreover, Sultan Selim saw the 'Mendes' family and its fortune as a potential asset for his empire, so he applied pressure on Venice to release 'Beatrice' and allow the entire 'Mendes' household to leave the city.

So when the 'Mendes' family made their entry into Istanbul in 1554, it was not just an arrival: it was a liberation. At last they could shrug off their Christian snake skins and live their true Jewish selves. 'Beatrice Mendes' became Gracia Nasi once more. 'Juan' transfigured back into Yosef Nasi. The family worshipped in one of the city's forty-four synagogues,[11] and Yosef had himself circumcised in April 1554. In June of the same year, he married his cousin Ana (ex Reyna) in a Jewish marriage ceremony.[12]

And Yosef, looking round for areas in which he could make money both for himself and for his new homeland, found alum.

And not only alum. The Ottomans farmed out the right to trade in important commodities such as alum, wine, wool, wax, spices, copper and leather to the highest bidders, and these regular payments became an important source of revenue for the Ottoman state.[13] Moreover, unlike

European rulers, they had no objection to Jews investing their knowhow and their money in these state monopolies,[14] so Yosef Nasi was soon managing the alum mines of Foça, while also running the wine trade with Moldavia (in present-day Romania), and the wine and wax trade with Poland. He was also involved in collecting taxes.[15] The income made the Nasi family one of the wealthiest in Istanbul, with a fortune of 400,000 ducats. Eighty people were regularly hosted at the Nasi dinner table in the *Belvedere*, their palace with a grand view over the Bosporus.[16]

Most Ottoman-produced alum was of course sold to dyers within the empire. But even though the Frescobaldi racket had come to an end soon after Henry VIII came to the throne in 1509, the Ottomans were still continuing to circumvent the pope's ban on Turkish alum and to sell alum in Europe. Much of this alum, which was of course illegal in the eyes of Rome, was traded via Venice, and indeed for three years from 1547, the Venetians themselves actually ran the Kütahya alum mines (now called the Gediz mines), producing 75 tons of alum a year.[17]

Yosef Nasi ran the alum mines of Foça along with a fellow Jewish magnate, Aharon de Segura, and very soon after taking over the mines, the two partners hit a problem. The Ottoman alum was sold in Venice through an intermediary called Hayim Saruq, who sold it on to the Italian merchants. And what should happen in August 1566, just after a huge cargo of Ottoman alum had been dispatched to Venice, but that Saruq went spectacularly bankrupt. His debts, according to his own account, amounted to a massive 112,000 ducats.[18]

What would now follow was obvious to all. The Venetian authorities would seize the valuable cargo of alum when it arrived in Venice, it would be considered an asset of Saruq's, and it would therefore be sold off in order to pay off his many Italian creditors, with a cut for the Venetian state, of course. Little, if any, of the money made from the sale would trickle back to the alum producers in Turkey, who would, in consequence, suffer an enormous loss. And 'enormous' is almost an understatement. At a time when the total annual production at Koloneia (now known as Sebinkarahissar in Turkish) was 357 tons of alum a year, the one shipment now making its way to Venice was 15,000 cantara, or 846 tons.[19]

But what could be done? The alum was already out at sea. The ship could not be contacted or intercepted. And indeed when it duly arrived in Venice, its valuable load was seized by the authorities. As far as they were concerned, this was a private business matter between Saruq and his creditors, to be settled in a court in Venice.

But they had underestimated the magnates Yosef Nasi and Aharon de Segura. The two powerful Jewish magnates whispered a strategy into the Sultan's ear ... following which the Sultan's personal envoy turned up in Venice with a story that turned the case on its head.

For, according to the envoy, the 846 tons of alum that were now in the temporary possession of the Venetian state had never belonged to Saruq at all. It was not even the property of the alum company. It was, on the contrary, the personal property of the Sultan. Every ton of it had been promised to him, claimed the envoy, but the alum had then been illegally spirited away to Venice without the Sultan's knowledge. The Sultan, concluded the envoy, was very grateful to the authorities in Venice for having apprehended the alum before it was sold on, and requested it now be returned to him as soon as possible.

The story was of course preposterous, and neither Saruq's creditors nor the judicial authorities in Venice would have any of it. This was a private business matter, they insisted. It had nothing to do with anyone else.

But of course the Sultan's intervention had raised the stakes. The affair had now acquired a political dimension, and the government of Venice was involved. Selling the alum in Venice, it was realised, might now lead to a diplomatic row between Venice and the Ottoman Empire, with repercussions that could not yet be foreseen. And Venice, it was reluctantly but generally agreed, could not run the risk of disregarding its most powerful and dangerous rival in the Mediterranean. So, although the Sultan's claim was clearly outrageous, a solution was found in which Yosef Nasi and Aharon de Segura received fair compensation for their alum after all.

The alum saga shows the lengths to which the Sultan would go in order to defend the interests of his Jewish friends, because their interests, of course, intertwined so closely with his own. But more than that, it also exposed the weakness of Venice in the face of Ottoman intimidation. This was indeed symptomatic of the changing power relationship between the two great rivals, soon to be demonstrated militarily when the Ottomans expelled the Venetians from Cyprus in 1573.[20]

Yosef Nasi died a rich man in 1579. It was, for him, the end of a long road that had begun in Spain and led via Antwerp and Venice to Istanbul.

And it was the end of the road too for the grand old alum mines of Foça. Exploited on an industrial scale for three hundred years, their stocks were dwindling, and an economic recession in the late 1500s, with knock-on slump in the critical cloth and leather industries, probably sounded their death knell. They closed, exhausted, by the end of the century.[21]

But Ottoman production continued at Gediz (Kütahya), Sebinkarahissar (Koloneia) and Gümülcine (Komotini) in Greece, with a modest annual production of 400 tons a year.[22] The Gediz and Sebinkarahissar mines were still active in 1680 when they were visited by the great Turkish traveller, Evliya Çelebi, who recorded two yet further uses of alum that we have not come across before.

When his horse was wounded in battle, Çelebi dressed its wound with egg, salt and alum.[23] And in Istanbul, Çelebi recorded forty workshops which produced the very best black ink for use in the all-important Islamic art of calligraphy, an ink that was particularly resistant to fading in sunlight. The ingredients? Soot, gum, gall and – of course! – alum.[24]

12

Alum, the Reformation and a Scam

The Reformation, of course, tore the church in western Europe in two. It also ripped families apart, setting parents, children, siblings and cousins against each other. And few feuds were as tortured as that between Elizabeth I and her cousin, Mary Stuart, Queen of Scots.

Different in character but united in bloodlines and destinies, the two cousins found themselves on opposite sides of the new religious divide and rivals for the throne of England. A set of circumstances had put Mary at the mercy of her older cousin Elizabeth, who kept her prisoner in various English castles for nineteen years. There was little for Mary to do during this time other than write letters. Many she wrote to Elizabeth, pleading for her release. Others were written to English Catholics. And some letters were written to plotters prepared to risk their necks in order to set Mary free and crown her queen of England.

Such letters needed to be discreet, of course. The usual practice was to use normal ink to write a bland letter that would arouse no suspicion, and then to fill the gaps between the words with secret messages written in invisible ink – of which there was a surprisingly wide choice, including orange, lemon or onion juice, or vinegar, or even urine. Depending on the method used, the script could not be seen until the paper was heated, or treated with a powder such as coal dust, or soaked in water.

Now, as it happens, we know exactly the kind of invisible ink preferred by Mary Queen of Scots, because she refers explicitly to it in a letter intercepted by Elizabeth's spy master, Francis Walsingham. Advising her envoy in Paris to use invisible ink when he writes back to her, Mary adds "Alum appears to me the best, or gall nuts."[1]

In the murky world of Renaissance and Tudor espionage, then, alum had found yet another use. Any message written with alum, ground to a powder and used with vinegar, remained invisible until the paper it was written on was laid in water.

Not that it did Mary Stuart any good. She was executed in 1587.

The use of alum as invisible ink for agents and double agents remained a minor side-line, of course, in comparison with the need for alum as a mordant in the cloth industry, and here the Reformation had caused a serious rift.

The alum producers – the mines at Tolfa that supplied about two-thirds of all the alum used in Europe, and the Mazarron alum mine in Spain that produced much of the rest – were in territories that remained solidly Catholic.[2] But the largest market for alum, the Netherlands, had split in two, with the southern provinces, corresponding to Belgium today, still ruled by Spain, and the northern provinces, which fought a surprisingly successful military campaign against their Spanish occupiers, constituting a *de facto* independent Protestant country from about 1570. And the second largest market, England, had also broken with Rome. Excepting the relatively brief reign of Henry VIII's older daughter, Mary Tudor, England was firmly in the Protestant camp.[3]

Protestant countries naturally did all they could to cut off their traditional monetary contributions to Rome. Tithe money was withheld, monastic dues dried up as convents and monasteries were closed, and no more payments were made on the appointment of new cardinals for the simple reason that no more cardinals were appointed. But alum was a problem. For in the absence of any other reliable sources of high-quality alum, and with imports from Turkey at best intermittent, Protestant countries had to carry on buying Roman alum as before. Despite their best efforts, therefore, they went on sending money to Rome.

And there was worse. Tolfa's virtual monopoly in alum gave the pope unwelcome leverage over Protestant countries, a weapon he could wield against them if he so wished. This was no idle threat. In 1533, when Henry VIII divorced his Spanish wife Catherine of Aragon without having secured the pope's permission beforehand, Pope Clement VII retaliated with an alum boycott.

And it hurt. Cardinal Wolsey, in charge of the state's accounts under Henry VIII, had closed down the Frescobaldi racket in order to curry favour with Rome. After all, he hoped to become pope himself one day. England therefore relied once more on Tolfa alum, and when Clement VII turned off the tap, England's dyers were left high and dry. Meagre supplies of alum smuggled in from the Netherlands were nowhere near enough, and the country's pre-eminent industry, its cloth industry, plunged into decline. In desperation, England's dyers resorted to second-rate mordants such as bark, gall or a sulphate compound known as copperas, but they all produced colours that were dull and faded fast. The colours' names – *rat's colour, horse flesh, pease porridge tawny* and *goosetwrd green* – say it all. The trade in coloured English cloth to Italy withered, and there was a run on England's currency.[4]

So short, indeed, was Henry VIII of funds that when imports of Tolfa alum resumed in the mid-1540s, he didn't have the cash to pay for them. Instead, he bartered it for lead which he had stolen from abbeys and monasteries at the Dissolution, so the alum trade was responsible for the lead stripped from English abbeys ironically ending up on roofs in Rome.[5]

Across Protestant countries, the hunt was now on to find new, non-Catholic sources of alum. But there was no Protestant big bang to match the dramatic discovery of Tolfa alum in 1462. Instead, local entrepreneurs tried extracting alum from a kind of rock that was available more widely than alunite: alum shale.

The first attempts had in fact begun before the Reformation, and not by coincidence, given the legendary skill and ingenuity of German miners, they had taken place in the German-speaking lands. Thus there were primitive alum works near Cologne and in Tyrol as early as the 1460s, and at Schwemsal, north of Leipzig, before the end of the century.[6] But these early experiments were very hit and miss. In Plauen in Saxony, mining engineers Franz Lingk and Martin Schilling were manufacturing alum from alum shale by 1542, but the alum shale discovered in Saalfeld in Thuringia in 1530 was not touched for a dozen years or more as nobody knew what to do with it.[7] Further west, alum works were reported in Oberkaufingen in Hessen in 1554.[8]

The early pioneers blended ideas derived from alchemy with advice gleaned from new technical handbooks such as *De Re Metallica*, published in 1556 by a German doctor called Georg Bauer. Bauer told his readers to roast the alum shale for about eleven hours in a furnace, to sprinkle water over big heaps of the ore two or three times a day for forty days, and then to boil the ore in large copper cauldrons. This, said Bauer, produced an alum solution that you poured into troughs to crystallize.[9] It sounds precise, but in practice nobody knew how to assess the alum content in the shale. There was a lot to learn, and the age-old knowhow about how to extract alum from alunite only took you so far.

One of the most successful early enterprises was in Bohemia, with alum mines in Görkau (Jirkov) in 1554 and in Komotau (Chomutov) in 1558.[10] The chief operator in the region, Christof von Gendorf, produced enough local alum to cause the Bohemian price of alum to drop. He was encouraged in his efforts by King Ferdinand who, on hearing that Gendorf needed prodigious amounts of human urine to manufacture his alum, bestowed upon him the doubtful privilege of a monopoly on all the human urine 'from all the towns' in Bohemia.[11]

Not all searches for alum were as successful. In 1546, Henry VIII threw money at a mining project in County Wexford in Ireland. But even though German workmen were brought in, the enterprise, in the words of Robert Turton, 'swallowed up money' and the works collapsed with a 'substantial loss' in February 1553. One estimate is that the Crown lost as much as £6000.[12]

A second attempt was then made on the tiny island of Lambay, 30 kilometres off Dublin. And fascinatingly the island's owner, the Archbishop of Dublin, granted the entrepreneur, the Secretary of State John Chaloner, the right to prospect for alum on Lambay in 1564 in exchange for the same promise that the Byzantine Emperor had obtained from the Zaccarias for exploiting the alum in Phocaea three hundred years before: Chaloner undertook to clear the seas of pirates.[13]

It is a reminder of just how dangerous sea travel was, whether in the Mediterranean or the Irish Sea. The big difference was that the Zaccarias were phenomenally successful, and John Chaloner was not. There was no more alum on Lambay than in County Waterford.

Even initially successful works were often short-lived because the quantities of alum produced were too small to cover the costs of extraction. King Ferdinand of Bohemia tried to promote Gendorf's locally produced alum by banning all imports of alum from Tolfa and Spain, but the fact that he renewed the ban in 1551, in 1554 and 1556 suggests that foreign supplies were in fact still coming in. And no wonder. Despite all the effort, no alum shale works could compete with Tolfa in terms of quality, or with Spanish alum in terms of reliability. These sources were therefore still dominant across Europe and achieved record output in the middle of the 16th century: 37,000 cantara (over 1700 tons) of alum per year in Tolfa, and 26,000 (over 1200 tons) in Spain.[14]

It is ironic, then, that the alum that would really boost a Protestant country on a massive scale was Tolfa alum itself.

Building on their experience with Agostini Chigi, the popes had by the mid-1500s found a canny way of using Tolfa alum to reduce their bad debts. Every few years, they awarded the running, or 'farming', of the Tolfa alum mines to one of their chief creditors. The creditor made a fortune by milking Tolfa for what it was worth for a fixed number of years; and in return, they paid the pope a hefty fee or wrote off one of his particularly bad debts. So it was win-win for both sides. And as the alum mines reverted to Rome at the end of the agreed term, the pope could then award the contract for the following few years to a different creditor and have another debt wiped out.

In 1553 Paul III had granted the running of the Tolfa mines to a Genoese tycoon called Bendinelli Sauli, and when that contract expired in 1565 Pope Pius V awarded the contract to another Genoese by the name of Tobias Pallavicino. The Pallavicini were seen as a particularly safe pair of hands, for Tobias and his brother Agostino were already shipping Tolfa alum into Genoa so profitably that they had each built themselves extravagant palaces on the proceeds of their business. Tobias was in fact known locally as the 'king of alum'.[15]

Indeed the whole Pallavicino clan was in the alum trade. Tobias's son Fabrizio coordinated the operation in Rome. Two other sons, Orazio and Alessandro, sold the alum in London and Antwerp, now the leading port for the Netherlands.[16] A nephew, Niccolo, managed the alum business in Spain, where the Pallavicini were one of the country's most favoured Genoese families.[17] And in key ports where there was no family member, such as at Rouen, the chief French port for alum imports, the family had agents in their pay.[18] To all appearances, therefore, the Pallavicini were the perfect family to take over the running of Roman alum.

And indeed the family showed its mettle when, just as Tobias had been granted the Tolfa contract, the Netherlands rose up against their Spanish occupiers, with England supporting the Dutch and Rome supporting Spain. It was a tricky moment, both economically and diplomatically, but the Pallavicini were unfazed. From Rome they maintained supplies of Tolfa alum to Spain and to Antwerp; and from Antwerp they continued to export alum to Protestant Holland and Protestant England. In other words, they pulled off the feat of selling alum to all sides of the conflict. And when the Spanish authorities tried to weaken England by banning all trade between Antwerp and London, the Pallavicini coolly shipped alum directly from Civitavecchia to England, by-passing Antwerp altogether. This of course was a stab in the back for the Pope's Spanish allies, but the Pallavicini spared the Pope's blushes by using Venetian ships, not Roman ones. In 1577 Orazio even commissioned English ships to fetch alum not only from Genoa but even from under the pope's nose in Civitavecchia.[19]

However, all was not as it seemed. For far from being a safe pair of hands, the Pallavicini were swindling the pope under his very eyes. And in a grand manner.

The Pallavicini had in fact found a way of playing the limited-term contract to their own advantage. Each year, in addition to quarrying, shipping and selling alum as per contract, they were spiriting away vast amounts of undeclared, unsold alum, and stockpiling it at secret depots across Europe. The pope knew nothing about this, and derived no income from these secret hoards.

The exact amounts are of course shrouded in mystery, as missing sums often are, but the scale was enormous. It can be guessed at from unexplained short-term rises in the import figures in ports where the chief importers were agents for the Pallavicini: in Marseille, for example, with annual alum imports normally about 2,000 pounds in the 1570s, there was a spike of over 22,000 pounds (ten times as much!) in the one year 1573-74. In Rouen only 1,100 pounds worth of alum was imported in 1575-76, but over 48,000 pounds the following year, and over 63,000 pounds in 1578.[20] In fact, it is generally assumed that over the course of their twelve-year contract, the Pallavicini stole and hid enough alum to supply the whole of Europe for six years![21] It was the greatest alum robbery in history and it went on wholly undetected for over a decade.

It only came to light after the Pallavicini's Tolfa contract came to an end in 1578. For when Gregory XIII awarded the alum contract to the Tobias's rivals, the Lombard Bernardo Olgiatti and the Florentine Francesco Ridolfi,[22] Tobias's family pounced. Just as the new managers were trying to get to grips with the complex task of running Tolfa, the Pallavicini ordered their agents to sell alum from their secret depots across Europe. Suddenly, vast amounts of the alum appeared as though from nowhere: Europe was awash with the stuff and of course, the price collapsed. Alum had fetched 33 shillings per cantara before the release of the secret stock; it fetched under 20 shillings afterwards. In England the price of alum fell by 50%.[23]

This was all part of the Pallavicini's plan. Having been raking in the profits for more than a decade, they could weather a temporary fall in income – unlike the new alum farmers, of course, who could not make money with alum at such a low price. Their business model had, overnight, become unworkable, and production levels in Tolfa fell as the Pallavicini's rivals suffered painful financial losses.

That triggered part two of the Pallavicini's plan. As production in Tolfa faltered, the Pallavicini effectively had a monopoly of high-quality alum in Europe, and they now ordered their agents to reverse gear and restrict the sales of alum once more. This forced the price back up, and by 1585 alum was back up to 32 shillings per cantara. With the Tolfa mines operating at half-cock, the Pallavicini made a killing selling their illegal stock of alum at this high price.[24]

The pope, of course, was livid. He had been sold down the river, and back up again. First he had been cheated of a good part of his income while the Pallavicini had run the Tolfa business, and now he had lost again when the price of alum collapsed. And in fact he was still losing because he was now getting no cut from the Pallavicini's sales of stolen alum. Enraged, the pope had Fabrizio Pallavicino arrested and, in an effort to find out where the vast stocks of alum were hidden, he probably had the poor man tortured.

To no effect. The Pallavicini were sitting on a colossal pile of valuable alum. They had minimal costs: their alum had already been already mined, it was ready to be sold. They had the contacts with alum merchants across Europe, and they could guarantee reliable deliveries. They could sell their alum to whoever they wished.

The only question now was: who would they sell their alum to?

Disposing of stolen goods is rarely an easy operation. But the Pallavicini came up with so brazen and devious a scheme that it changed the face of European history and made one member of the family immensely rich.

At this point in the story, the focus changes from Rome to London, and from Tobias to his son Orazio Pallavicino. Or indeed Horatio Palavicino. In England Orazio's forename gained an H, his surname lost an L, and the man's career took off.

Life was often not that easy for Italian and Spanish merchants based in England in the period between the death of Catholic Queen Mary in 1558 and her husband sending the Armada thirty years later. There was fear of invasion, and foreigners were often suspected of being spies. And indeed some whispered that Horatio Palavicino was a Catholic agent, for all that he claimed to have entered the Church of England. Others said he was a *marrano,* a Jew who had outwardly converted to Christianity.[25]

Such allegations could, in Elizabethan England, cost a man his life. Elizabeth's physician Roderigo Lopez, a wealthy Catholic immigrant with a Jewish background, would face a trumped-up charge of treason and be hanged, drawn and quartered for a crime of which he was almost certainly not guilty.[26]

But Horatio seemed to rise effortlessly to the top. He became Elizabeth I's personal banker and the queen knighted Horatio in 1590. Elizabeth even wrote a rare letter appealing to the pope to release his brother Fabrizio from incarceration in Rome.[27] Horatio's meteoric career, so unusual in a foreigner at Elizabeth's court,[28] was of course not unconnected with the family's fabulous wealth, its network of valuable connections and its vast store of stolen alum.

Horatio had first tried, in 1578, to sell the hoarded alum to England, but Elizabeth's Secretary William Cecil, cautious of angering Spain because he feared a Spanish invasion, had vetoed the idea. So a second plan formed in Horatio's mind. An ingenious, complicated three-way deal in which the stolen alum would be sold to the Netherlands, but paid for by England. As Elizabeth's banker, he knew that the English state had the necessary funds, and knew exactly how Elizabeth could be deceived into paying up.

In the first part of the deal, Horatio proposed to sell £29,000 worth of the stolen alum to the provisional Dutch government, known as the States-General, with the alum supplied by Pallavicini agents in Antwerp and Genoa.[29] The States-General would be able to sell the alum to Dutch dyers and use the proceeds to buy much-needed military hardware for their war against Spain. In return, the States-General would guarantee the Pallavicini a six-year monopoly on the supply of alum to the Netherlands.

Of course, Horatio knew very well that there was not the remotest chance of the cash-strapped States-General, up to their necks in the expense of war against Spain, having the means to pay for such an enormous amount of alum. So the second part of the deal was a financing plan. Instead of paying for the alum up front, the Dutch could take the £29,000 as a loan, to be paid back later. The loan would be underwritten by Elizabeth I of England as a show of support for fellow Protestants in their fight against a common Catholic enemy. In practice, of course, everyone knew that the Dutch, being bankrupt, would default on the loan, so Elizabeth would be liable for the payment.

Not, Horatio hastened to assure Elizabeth, that she need ever pay so large a sum. For the third part of the plan was that Horatio would himself lend Elizabeth the £29,000. And Elizabeth need not pay it back to him if she did not wish to do so. All she would have to do is pay Horatio the interest on this loan to her.

The deal, signed on 10th July 1578,[30] was of course a godsend for the Dutch rebels, who received £29,000 worth of alum without having to pay anything for it. And it appealed to Elizabeth, who liked to be seen helping her Dutch allies, especially if, as her banker assured her, she would never be required to re-pay the sum she had borrowed.

And Horatio? On the face of it he would be £29,000 out of pocket. But only if you ignore the interest payments. For remember, the deal tied Elizabeth into paying annual interest payments to Horatio. Now, as recently as 1571 Elizabeth's government had re-introduced a statute of Henry VIII's of 1545 which allowed interest rates of up to 10%.[31] Take the time factor into consideration and Horatio stood to gain more in accumulated interest payments than the value of the alum loaned. And Elizabeth would still owe him £29,000 at the end of it.

The one factor missing from the deal was, of course, any hint of ethical or moral principle. In this respect, the Pallavicini were typical of Europe's Renaissance banking families. Families like the Welser and Fugger in Augsburg, the Medici in Florence and the Greshams in England understood the power of money, the value of securities and the role of interest at a time when the aristocrats they served usually considered it beneath their dignity to concern themselves with such matters. So they made money: morality did not come into it. They had no more qualms about exploiting their sovereigns' ignorance of finance than about investing in Europe's trade of slaves from Africa to the Americas.

⁂

The consequences of Horatio's deal were far-reaching.

Maybe most importantly, it kept the Dutch insurgents afloat. They had been on their knees for lack of money while Spain, Europe's superpower in the 16th century, was benefitting from the gold imported from South America,[32] and this imbalance might well have led to a Dutch defeat. Instead, Horatio's alum deal enabled the rebels to pay their soldiers and purchase new weapons, and therefore helped to secure the ultimate success of the Dutch over the Spanish. The Spanish Netherlands became a hugely significant bastion of Protestantism in continental Europe at a time when Protestants in France and Hungary were suffering painful defeats.

The deal also ensured that the all-important Dutch cloth industry always had plentiful stocks of alum. Military actions such as the sacking of Antwerp by the French in 1583, and the blockading of Dunkirk by Spanish troops in the same year, might under other circumstances have left the Dutch dyers with no access to alum,[33] but thanks to Horatio's deal the Dutch cloth industry flourished, paving the way for the economic boom which would transform the Netherlands into Europe's economic powerhouse.

What's more, the Dutch economic triumph was in painful contrast to the troubles in Spain, where Horatio's manipulation of the alum market had devastated the alum industry at Mazarron in Castile, and contributed to its terminal decline. Prior to 1578, production in the three works in the Mazarron area had hit 26,000 cantara (over 1,200 tons) a year, but Horatio's deal, coming on top of the effects of the Reformation and of war, caused its sales to shrivel to 3,000 cantara (142 tons) a year in 1588, and in 1592 Mazarron closed for good.[34]

Rome and Spain were enraged, of course. The Pope had been swindled, and a valuable Spanish industry was being decimated. So they tried to hit back. But a Spanish order in 1581 to seize any vessel in the service of the Pallavicini was hard to put into practice.[35] A full alum boycott, like that imposed on Henry VIII by Pope Clement VII, would have hurt much more, but the abundance of the Pallavicini's stolen alum made this impossible. In frustration, Horatio was tried in Rome by the Inquisition, was condemned in absentia in 1584 and all his Italian property was seized. Effigies of Horatio were burned in Genoa and Rome.[36]

For Horatio, this was water off a duck's back. He was so wildly successful in England that he could easily do without his property in Italy. For his alum gamble paid off royally: the family's stock of stolen alum was disposed of in less than ten years,[37] and Elizabeth continued to make interest payments to him for another five years after that before she realised that she was paying through her nose for nothing at all. By the time she cut the payments in 1593, Horatio had been taking interest payments for fifteen years, earning him a colossal £45,479 – about 50% more than the value of the alum, which had of course been stolen in the first place.[38] The Pallavicini had outwitted the pope and the Spaniards, and now they had outwitted England's Virgin Queen.

Elizabethan aristocrats were not paid a salary for their services to the state. However, their expenses were enormous, because they were expected to pay royally for elaborate food, clothes, and festivities whenever Elizabeth chose to stay in one of their residences. Many too, lost eye-watering amounts of money gambling. As a result, many influential aristocrats lived well beyond their means and died virtually penniless or in debt.[39] Francis Walsingham, Elizabeth's master of espionage, for example, had such large debts that he was forced to request to be buried 'without any such extraordinary ceremonies as usually appertain to a man serving in my place'.[40]

Not Horatio Palavicino. Thanks to his alum deal, he died in July 1600 one of the richest men in England, leaving an estate valued at £100,000.

Plus, of course, the £29,000 that Elizabeth still owed him – which, as far as I have been able to make out, was never paid back.

13

Alum in Yorkshire

I am standing on the west bank of the river Esk, high above the little town of Whitby. To my left, Whitby's two piers stretch their protective arms out into the open sea, and the tidal harbour is snug in the narrow valley in front of me. Fishing boats bob on the sheltered waters by the quays, cosy fishermen's cottages with their orange-red pantile roofs line the narrow streets. High on the cliffs opposite me the stumpy parish church and the majestic ruins of Whitby Abbey give their blessing to it all.

Standing here on day trips from York I have often told visiting friends and family the standard script about how Whitby grew from a tiny fishing harbour into a port of national importance on the back of its whaling industry. Cut off from the rest of the country, so the story goes, by the boggy, foggy, soggy North York Moors, with nowhere to turn to but the sea, the men of Whitby ventured far into the oceans to catch whales. And up here high on the west bank, this narrative feels very plausible. Next to me is a statue of Captain James Cook, on whose expeditions the future whalers of Whitby became acquainted with the southern and northern seas. Below in the harbour is a replica of the *Endeavour*, the sturdy, flat-bottomed Whitby-built ship on which they sailed. And my very view of the harbour is framed by an arch consisting of a large, upright, single jawbone of a whale.

Even so, there is (please excuse the pun) something fishy about this story. Whitby's whaling industry took off in the eighteenth century, but the town's population had leapt from under 200 in 1540 to 3,000 by 1660. Whitby's primitive wooden piers had been replaced by sturdier stone piers in 1632. And while there were three trading vessels in 1540, there were sixty in 1690.[1] Nor did James Cook learn his seafaring skills on fishing boats. Every biography insists he trained in what is mysteriously called 'the Whitby coal trade'. Coal trade? Whitby had no coal to export, and its modest population had no need to import coal on any significant scale. Why a Whitby coal trade?

The answer lies in an earlier, even smellier trade than whaling, an industry not commemorated in any statue. What transformed Whitby from an insignificant fishing harbour into a busy industrial port was the alum industry. Alum shale was quarried near Whitby from the early seventeenth century, and huge amounts of coal were needed to burn and process it. In addition, urine and later kelp, or seaweed, were needed in the process of converting the burnt shale into useable alum. It was the modest coastal vessels that brought these materials, and which exported the alum itself, that gave Cook and the men of Whitby their first experience of managing ships in the open sea. As Lionel Charlton, a maths teacher in Whitby, put it in 1779, the alum trade 'raised us out of obscurity, made us acquainted with navigation, and has rendered us of such consequence as a maritime town'.[2]

Indeed, Yorkshire's alum industry kick-started the shipbuilding industry that built Cook's first ship. It was because so many boats were needed for the alum industry that it became worthwhile to build ships in Whitby. And as Whitby's harbour was tidal, with no system of locks to trap water at high tide, its shipwrights specialised in building sturdy, flat-bottomed boats that easily settled on the mud at low tide and re-floated as the water rose again.

From his experience on such Whitby Cats, as these boats were known, James Cook knew they were ideal for sailing in unknown waters, for they could easily be beached and repaired, with no need to find a deep-sea harbour. It is no surprise, therefore, to discover that the *Endeavour*, the ship that Cook chose for his first long voyage that began in 1768, started life as the *Earl of Pembroke*, a flat-bottomed collier, Whitby-built to service the alum industry.

⬩⁘⬩

Back in the early 17th century, the track record for finding alum in the British Isles wasn't great.

Henry VIII's attempts in Waterford had been as fruitless as John Chaloner's on the island of Lambay. Under Queen Mary's Catholic regime, there were no objections to importing alum from Tolfa, but when searches for alum resumed under Elizabeth I, this time in the south of England, they met with the same lack of success.

On claiming to have found alum on the Isle of Wight in 1552, William Kendall was granted a monopoly on alum mined in England for twenty years. He backed out after just two. Cornelius de Vos from Liège was then awarded the alum monopoly for twenty-one years, but he too ran out of money, and he sold out to James Blount, 6th Lord Mountjoy in 1565 or 1567. Mountjoy then promised to produce 150 tons of alum in Dorset within two years if the Queen advanced him the sum of £6,000, but the Queen wisely kept her money, and instead granted Mountjoy a monopoly on alum for twenty-one years, while reserving ten per cent of the profits for herself. She did not need to hold her breath. But at least Elizabeth did not lose her money, unlike investors who had been induced to put money into the scheme. Despite the long and tedious legal squabbles about who should own the alum, were any alum to be found, no alum was found, and none produced.[3]

Alum was never going to be found in the south of England because the geology was wrong. In Yorkshire, in contrast, a stratum of alum-bearing shale had been lifted by a geological fault that cuts across the northern edges of the North York Moors and curves for fifteen miles along the coast to Whitby and beyond. So here, at last, the search for alum was crowned with success. Strangely, though, we do not know for sure who made the first successful dig.

The most likely contender is John Atherton, who in 1603 discovered alum on his land at Slapewath, two miles east of Guisborough on the northern edge of the North York Moors, and in 1604 set up a processing site in partnership with Sir John Bourchier.[4] Intriguingly though, a plot of land at Stoup Brow near Ravenscar was called Alum Hill in a document as early as 1538.[5] Further, a Lord Danvers claimed an interest that may or may not have been associated with alum works in Grosmont, held by two different historians to be the earliest in the region when they visited their ruins in 1801 and 1817.[6]

The most frequently quoted version of events, however, is the wonderfully gung-ho story of one Thomas Chaloner, often described as the tutor to the Prince of Wales, who noticed one fine day in 1604 that plants growing on his estate at Belman Bank near Guisborough were similar to some he had seen while travelling near the pope's alum works in Tolfa. 'Aha!' he said to himself, 'there must be alum here!', and indeed when he began to dig he found alum-bearing shale. He did not, however, know how to process the shale, and the pope had bound his workers in Tolfa to secrecy. So with Drake-like panache, Chaloner swooped into Civitavecchia, bribed some Italian alum workers to accompany him, and sailed away with them hidden in a barrel on his ship. With this clever trick, foreign obstruction was overcome, the English alum industry was born, and an irate pope was left to lay a useless curse on the intrepid English adventurer.

Alas, little in this flag-waving tale is true. Nor is it even very original, for identifying new alum deposits by associating certain plants with an existing alum site is of course a repeat of how Castro is said to have discovered the alum mines at Tolfa. For the record, no such plants have ever been found in Yorkshire, nor has any evidence for the kidnapping of the Tolfa workers. The workers would in any case have been of little use as their expertise was in extracting alum from alunite, not from alum shale. What's more, the Thomas Chaloner who is supposed to have discovered alum in Yorkshire was almost certainly not the Prince of Wales' tutor who owned Belman Bank, but the tutor's cousin – the son of John Chaloner who had prospected for alum on the island of Lambay, and who was also called Thomas. And as for the papal curse, this was an embellishment added to the tale some hundred and sixty years later, and says more about anti-Catholic sentiment in the eighteenth century than it does about the discovery of alum.[7]

True or not, however, the story of Chaloner's discovery of alum inspired three heavyweights at court – Edmund Lord Sheffield, Sir David Foulis and Sir Thomas Chaloner (the owner of Belman Bank and cousin of the illustrious discoverer of alum) – to champion a new Yorkshire Alum Company based at Belman Bank. They silenced Lord Danvers's rival claim by paying him an annuity 'for unstated reasons'[8] and *de facto* acknowledged the prior claim of the works at Slapewath by inviting Sir John Bourchier to join them at the top table. And by pulling the right strings at court they secured a Patent from King James I in 1606 (amended in 1607 to accommodate Sir John Bourchier) giving them the go-ahead for Yorkshire alum production.

The King awarded the company all rights on English alum for 31 years in return for which Sheffield, Foulis, Chaloner and Bourchier, now known as the *patentees*, promised to produce 1800 tons of alum per year, equal to the annual consumption of alum in England.[9] The *patentees* did not have to put up any of their own money, but their wealth guaranteed loans to the company by investors, such as London merchants.[10] Capital was also put up by the five *farmers* to whom the patentees leased the works. The *farmers* in turn employed contractors, or *undertakers,* to manage the day-to-day functioning of the works. Profits were to be shared between patentees and farmers according to an agreed formula, and the patentees also committed to pay the King £700 a year, equal to the amount of customs duty charged on current imports of alum from Tolfa and Spain. Imports would of course no longer be necessary once the Yorkshire alum works hit their production targets.[11]

It sounded great, but was not. The figures had been plucked from thin air. Nobody knew whether the Yorkshire alum reserves could yield 1800 tons a year or not, nor did anyone have the remotest idea how much this alum would cost to produce. In fact, nobody had much experience in manufacturing alum from the shale, which explains why a small number of workers were brought in from Liège, where the shale-alum industry had got under way in 1586.[12] With so many unknowns, the company not surprisingly produced more losses than alum: losses of £20,000 in the first year and £30,000 in the second (some £400,000 at 2018 prices).[13]

Thus the scene was set for the story of alum in Yorkshire in the following fifty years. An impressive industrial enterprise, on a scale never seen before in England, would employ hundreds of people, remove thousands of tons of rock and change the very outline of the Yorkshire coast. But this brave new world of industry would be compromised by the complicated, ethically doubtful and often downright corrupt financial shenanigans of a group of privileged 17th-century gentlemen whose only interest was to make easy money at someone else's expense.

Alum was mined and manufactured for over two hundred and fifty years at more than twenty different sites in north Yorkshire, in a great arc from Thimbleby and Carlton Bank on the western edge of the Moors, through Guisborough and Slapewath to the north, on to Loftus, Boulby and Sandsend on the coast north of Whitby, and down to the Peak and Stoup Brow works at Ravenscar. There were works in the Esk valley too, at Littlebeck and Godeland Bank.

Each site produced alum from shale in a laborious industrial operation consisting of many different processes whose fundamentals changed remarkably little over time, so that Turton's description of the process in 1612 does not differ in essentials from Walter White's description of an alum works in 1861 (and readers who would like to see a full chemical description should consult Peter Appleton's wonderful book about the Yorkshire alum industry, *A Forgotten Industry)*.[14] But visitors in the early seventeenth century will certainly have been stunned by the scale of the enterprise.

The first sign of the works will have been pillars of sulphuric smoke rising into the sky, and visitors will then have been assailed by acrid smoke and ash that filled the air. The smoke billowed from vast heaps of smouldering shale, known as *clamps*, that could be up to thirty metres wide at their base. Layers of shale and brushwood were laid alternately on them, and set alight when the clamp was something over a metre high. More and more shale and brushwood were then heaped onto the burning pile, till the clamps were anywhere between fifteen and thirty metres high. Taking anything up to eight or nine months to build, they were usually sited next to a cliff edge so that, from a certain stage, the shale could be tipped from above. For the labourers, of course, this meant getting perilously close to the edge, working in clouds of sulphurous white smoke which got worse as the burning worked its way up though the clamp. The scene, wrote White, had 'the appearance of a volcano' whose fumes gave him 'a painful sense of choking, with an impulse to run'.[15]

The shale was brought by barrow from the quarry or mine. First, though, a layer of earth, stones and rock, fifteen metres thick or more had to be removed by pick and shovel before the shale could be reached.[16]

The shale itself was, according to Walter White 'like slate soaked in grease: it has a greasy or soapy feel between the fingers'. Being softer than the rock above it, it was easier to mine and the labourers broke it into pieces small enough to be wheel-barrowed on wood or iron barrow ways to the clamps.

The clamps burned for three or four months or more, with great skill required to regulate the burning, or *calcining*, of the shale. If the fires burned too fiercely, the flames consumed the wood before the shale was broken down, so to dampen the flames, moist earth was plastered to the sides of the mounds. Too much earth, however, risked putting out the fires altogether. If all the shale had fully burned, fifty tons of shale would yield one ton of alum.

Once it had cooled, the burnt shale was dug out of the clamp, loaded onto barrows or ox-drawn wagons, and taken to large *steeping* pits, some eleven metres long, four and a half metres wide and a metre deep, that had been hewn out of the rocky ground. Each was large enough, according to White, to accommodate forty wagon loads of burnt shale which was left to soak in the pits for three days, to allow soluble chemical compounds to leach out into the water.

The steeping pits were lined with stone, and stone was also needed to build the alum houses and the furnaces. Stone was available locally, of course, but nearby was a ready source of stones that had already been cut: the empty stone shell of Whitby Abbey which, in the 1600s, was still virtually complete. Turton notes that it was established practice to collect stone from the 'old Abbey', leading to the collapse of the south transept in 1736 and of the abbey tower in 1763. We therefore have the alum industry, as well as local house builders, to thank for the depleted ruins of Whitby Abbey today.[17]

The five men working at each steeping pit stirred the murky liquid till they judged it had reached sufficient concentration, at which point it was called the *liquor*. None of the labourers, supervisors or managers understood the chemical processes that took place in the clamps and steeping pits; instead, they proceeded by trial and error, and sometimes, inevitably, they made mistakes. Dr Edward Jordan, who took over the Slapewath works in March 1614, believed he could produce alum more cheaply by using peat instead of coal in the furnaces, and chalk instead

Alum crystal
This is an alum crystal at the end of the manufacturing process. Crucially for its centuries-long role as a mordant, it is soluble in water. Today it is used to as a pickling agent, as a deodorant, and in styptic pencils to staunch minor bleeds. (Wikimedia Commons)

Kharga oasis
Set in the desert west of the Nile, Kharga oasis was an important staging-post on the trans-Sahara Darb El Arba'in trading route. Alum was available here as an evaporite on the edge of the shrinking oasis lake. Later it had to be mined. (Wikimedia Commons)

Alum in the decorative arts
Nearly four thousand years ago. Egyptian craftsmen laid figures in a solution of alum, copper alloys and vinegar to give them a dark sheen, or *patina*, which contrasted with areas of polished gold or silver. This crocodile in the Museum of Egyptian Art in Munich was made in about 1800 BCE. (Wikimedia Commons)

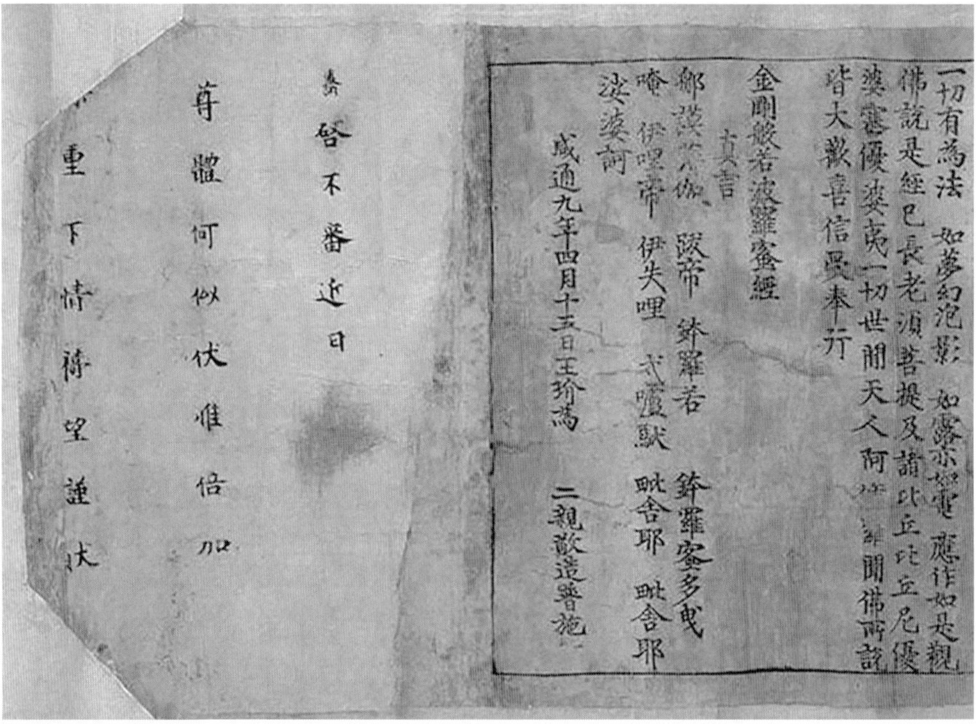

一切有為法 如夢幻泡影 如露亦如電 應作如是觀
佛說是經已長老須菩提及諸比丘比丘尼優
婆塞優婆夷一切世間天人阿修羅聞佛所說
皆大歡喜信受奉行
金剛般若波羅蜜經
真言
那謨婆伽 跋帝 鉢羅若 鉢羅蜜多曳
唵 伊哩帝 伊失哩 戍嚧馱 毗舍耶 毗舍耶
娑婆訶
咸通九年四月十五日王玠為 二親敬造普施

The last page of the Diamond Sutra
China was the first country to make paper, and alum was used in the manufacture of top-quality paper that could stand up to the wear caused by woodblock printing. Printed in 868 AD, the Diamond Sutra in the British Library is, it is claimed, the oldest printed book in the world. (Wikimedia Commons)

The Leiden papyrus
Written in the third century AD, this papyrus describes how to use alum to remove copper and other metals from the surface of objects made from debased gold, in order to suggest that the items are made of pure gold. Discovered in Thebes, on the Nile, the papyrus is named after the city in the Netherlands where it can now be seen. (Wikimedia Commons)

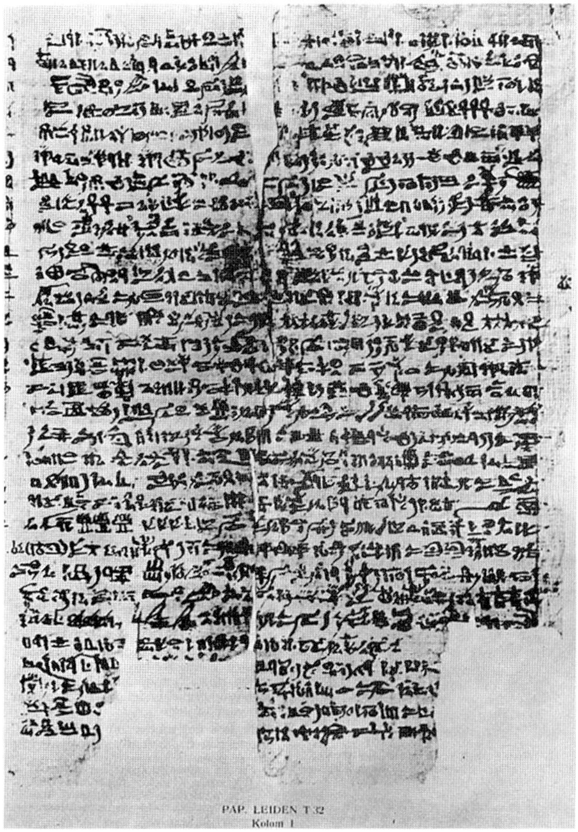

Gold coin from Aksum, Ethiopia
Alum made it possible to enrich the surface of coins made from debased gold or silver, thereby making them look as if they were made of the pure metal. Starting in Egypt, this practice spread within the Roman Empire and beyond – to Aksum in present-day Ethiopia, for example, where this fourth-century coin was in use. (Wikimedia Commons)

The walls of Byzantium
By the 10th century AD, the Byzantine Empire had shrunk to an area covering not much more than Greece, Macedonia and part of present-day Turkey. The fact that it managed to survive for another five hundred years is significantly down to the extraordinary wealth it derived from its alum mines. Ensconced behind their massive walls, the Byzantines could afford to buy off their enemies. (Wikimedia Commons)

Phocaea (Foça)
The Genoese were awarded the huge alum mines of Phocaea in return for supporting the Byzantine Emperor Michael VIII Palaiologos against the Venetians. From here they initiated the long-distance trade route to Bruges and Southampton. The Genoese castle was largely restored by the Ottomans. (Wikimedia Commons)

A medieval Italian galley
In 1295 the Genoese had 165 galleys for use in trade and war. It was hard to fill a galley with exotic spices and fruits. But filling the holds with alum and then topping up the load with exotic goods from the East made trading extremely profitable. The Genoese fleet became, in essence, an alum fleet. (Wikimedia Commons)

***Left*: Bruges cloth hall**
The beautiful belfry of the Cloth Hall in Bruges, re-built in 1280 after a fire, still dominates the city. The octagonal top was added in the 15th century, when a booming cloth trade made Bruges one of the wealthiest cities in Europe. Alum from Phocaea was an essential ingredient for the cloth trade. (Wikimedia Commons)

***Below*: Cloth Hall, Southampton**
Southampton's cloth hall is a much more modest building than its equivalent in Bruges. Now called Westgate Hall, it was built in about 1400. Cloth was traded upstairs, and fish downstairs in an open arcade. The building originally stood in St Michael's Square. (Wikimedia Commons)

Caffa (Feodosia)
Genoa's trading post in Crimea was a large city surrounded by walls that were five kilometres long. It had shipbuilding yards, a mint, a castle, a mosque, two synagogues and Catholic, Orthodox and Armenian churches. Spiced with a hint of danger because it was so far from Genoa, it was a place where enterprising young men could make a fortune – in the alum trade, for example. (Wikimedia Commons)

Trebizond (Trabzon)
High quality alum from Koloneia came overland to the Byzantine statelet of Trebizond on the Anatolian (Turkish) Black Sea coast, from where much of it was brought to the Genoese trading post of Caffa. Trebizond was independent of, and outlived, the Byzantine Empire in Constantinople, only falling to the Ottomans in 1473. (Wikimedia Commons)

The castle in Sinope
Castles around the Black Sea, from Romania to Georgia, are called 'Genoese', although most were later rebuilt or re-modelled by the Ottomans. Sinope on the Anatolian (Turkish) Black Sea coast was an alternative outlet to Trebizond for alum from Koloneia. (Wikimedia Commons)

Sultan Mehmet II
In May 1453, Ottoman forces under Mehmet II took Constantinople. Christian powers hoped his occupation would be of short duration, but Mehmet dashed their hopes by mopping up all remaining independent territories in the region, starting with the alum mines of Phocaea in 1455. (Wikimedia Commons)

***Above left*: Madder**
Red dyes made from madder root, fixed to the cloth with alum, were used in Egypt as early as 1500 BCE, and maybe earlier. They continued to be used into the Middle Ages, when even brighter reds could be achieved with vermilion and cochineal. (Wikimedia Commons)

***Above right*: Woad**
Blue dyes were made from woad in a particularly smelly process which involved leaving the leaves to ferment for nine weeks. (Wikimedia Commons)

Murex mollusc
The decomposing molluscs secreted an evil-smelling yellowish liquid from which a deep purple dye could be obtained. Dyers were looked down on because their trade gave off such unpleasant smells. (Purple dye murex or spiny dye-murex, *Bolinus brandaris*, location: Italy, Porto Garibaldi. (Wikimedia Commons)

Parti-colours for the party?

As dyers produced ever more striking colours, both men and women in rich families competed in showing off as many colours as possible in their clothes, shoes, belts and hats. In parti-coloured clothing, the colours on the left were different from colours on the right. All the colours were made possible thanks to the trade in alum. (Wikimedia Commons)

***Above left*: Pope Pius II**

The pope was delighted when vast reserves of alum were discovered at Tolfa, just outside Rome. The income from the sales of alum were, in theory, to be used to fund a campaign to regain Constantinople, but Pius used some of it to revamp his home town which he named Pienza, in honour of himself. (Wikimedia Commons)

***Above right*: Pope Leo X**

Alum provided the popes with a huge income, but also tempted them to spend more lavishly. When Pope Leo's ruinous war against Urbino depleted the income from alum, he had to resort once more to the sale of indulgences. (Wikimedia Commons)

Agostini Chigi

A 12-year contract to run the Tolfa alum mines made Chigi one of the richest men of his age and he was renowned for throwing sumptuous dinner parties. Unlike the popes, however, he made sure his outgoings never exceeded his income. (Wikimedia Commons)

***Above left*: Sultan Selim II**

While Spain drove out its Jewish citizens, Selim made Istanbul a safe haven for Jewish people. Yosef Nasi, a Jewish entrepreneur, ran the mines in Phocaea, producing alum for use within the Ottoman Empire, and, despite the papal ban, for export to Europe via Venice. (Wikimedia Commons)

***Above right*: Tobias Pallavicino's palace in Genoa**

Well, one of them. Appointed by Pope Paul III to run the Tofa alum mines in 1565, Tobias Pallavicino defrauded the pope on a massive scale by hiding vast stocks of alum. Tobias's son Horatio then made money out of Queen Elizabeth I by charging interest on money he lent her to pay for the stolen alum. (Wikimedia Commons)

The former alum mine in Saalfeld

After the Reformation, Protestant countries searched for alum in order to free themselves from buying Tolfa alum. Miners in the German-speaking countries first found out how to obtain alum from alum shale. Several former alum mines in Germany are now open to the public. (Wikimedia Commons)

Georg Bauer's *De Re Metallica*

Bauer's technical handbook *De Re Metallica,* which had a section on alum mining, became a standard work across Europe. This woodcut illustration showed the process of producing alum. Bauer often went by the latinised version of his name, Agricola. (Wikimedia Commons)

The Yorkshire alum works
Shale was hewn from cliffs, transported, burned, steeped in water and then repeatedly boiled in a chemical industry that began in 1604 and preceded the Industrial Revolution by more than 150 years. As for the alum workers, they were often paid months in arrears. (Wikimedia Commons)

The former alum works in Sandsend
The buildings associated with the Yorkshire alum industry have largely disappeared but the long-term effects of the industry are clearly seen in the headlands at Kettleness, Sandsend and other places where the alum was once worked. A thick layer of discarded alum slag has created a barren, moon-like surface that contrasts with the grass in the foreground. (Wikimedia Commons)

Right: **A 18th century chintz dress**
Brightly coloured Indian cotton fabrics took Europe by storm in the late 17th and 18th centuries. Iron and alum, thickened with resin, were used as mordants. Unlike dyers in Europe, Indian dyers had found a way of removing the resin after printing the colours, and this made the fabrics brighter than those manufactured in Europe. (Wikimedia Commons)

Below: **Christina Piper's castle near the Andrarum alum works**
Under Christina Piper in the 18th century, the Andrarum alum works became one of the largest industrial enterprises in Scandinavia. Christina ran a tight ship: the works included a school, a pharmacy and a hospital, but also a court house, a prison and a flogging post. The ruins of the works can be visited today. (Wikimedia Commons)

The alum works at Hurlet, Glasgow
Charles Macintosh, who would later go on to invent the waterproof raincoat, developed a more efficient way of manufacturing alum by boiling the concentrate in a boiler furnace, rather than in open pans as in Yorkshire. This led to a fall in the price of alum and was the beginning of the end of the Yorkshire alum industry. (Wikimedia Commons)

The Galata tower in Istanbul
The Galata Tower, built by the Genoese in in 1348, is an icon of modern Istanbul. It is a reminder of the Genoese trading empire in the eastern Mediterranean and Black Sea, an empire that was largely built on alum. (Wikimedia Commons)

of urine as an alkali. But the experiment failed and Dr Jordan sold out at a loss before the end of the year.[18]

When the liquor was judged to be strong enough, it was run off in a conduit to the alum house. But the burnt shale was soaked two or three times again so as not to waste any alum within it. Only after the final soaking did the men remove the sludge that remained in the pit, making space for the next load of alum ore.

In the alum house, meanwhile, the liquor was boiled in large lead pans that were suspended over coal fires. They burned – and smoked – all day. The coal came from the alum company's mines at Harraton, on the River Wear near Sunderland, and was shipped to Whitby for the coastal works and, as Middlesbrough did not as yet exist, to Coatham on the Tees estuary for the Guisborough mines.[19]

The liquor in the pans grew in strength as the liquid evaporated, and when sufficient specific gravity was achieved, the liquor passed to a 'settler' where, in modern parlance, an alkali was added to cause the precipitation of the alum crystals.[20] Not that the 17th-century alum workers would have understood such terminology. As far as they were concerned, you added urine. It was sourced locally at first, with the urine of working people, who consumed 'little strong drink', considered superior to that of the gentry. Not that this prevented Sir John Chaloner (of Lambay) from contributing his urine at a penny per forty litres. However, people soon realised that they could earn more money if they diluted their urine with water, and this made it less effective in the alum-making process.[21]

As alum production increased, urine had to be shipped in from Hull (where, it is claimed, the first public urinals were built in order to collect urine for the alum works)[22] and even from London, on boats that brought the alum to London and would otherwise return empty. Colossal amounts of urine were required; in November and December 1612, for example, the alum works at Sandsend took delivery of 13,000 gallons of urine. We therefore have the alum trade to thank for the expression 'taking the piss': Captain Luke Foxe, master of the *Allomes Amye* who, having delivered 29 tonnes of alum to London, was returning to Whitby with 23 tons of urine, was doing just that.[23]

It was, as we have seen, when urine (and, later, kelp) was added to the solution that the valuable alum crystals finally formed. But even then the end of the process had not yet been reached, for the liquor above the crystals (now called the 'mother liquor') was removed to be re-used, while the alum crystals were washed, dissolved and boiled again in a minimum of water to increase the alum content. The alum makers had to judge when the alum solution was strong enough, and there is a quaint tradition that they knew the liquor had achieved the ideal concentration when they could float a raw egg on it. A wonderful story, but there is no evidence for it.

When the desired concentration had been reached, the alum solution was pumped into large wooden 'roaching' casks nearly three metres high, and left for two weeks or more to allow the alum solution to crystallise. And it was when these casks were taken apart that solid, cask-shaped, three-metre-high lumps of white alum finally emerged. Impressed, Walter White was moved to describe the whole process as 'a happy triumph of chemical art'. And indeed this was more than mere manufacture. It was a process in which a product was chemically purified, and therefore England's very first chemical industry.[24]

The immense scale of the operation can be judged from the fact that by the 1800s Yorkshire was producing 3,000 tons of alum each year. Each of those tons required thirty-six tonnes of surface rock to be removed, up to fifty tonnes of shale to be mined, barrowed and burned, six tons of coal to be brought and burned, and over 500 litres of urine (or one ton of kelp) to be added. And all of this work was done by hand.[25]

The works needed diggers, quarrymen, miners, labourers, supervisors, bricklayers, carpenters, coopers, ironsmiths, carters and more. There were the men who led the alum-laden donkeys along the stone-slabbed donkey trods that connected the moors to the coast, the crews that worked the boats and the people who sold food to the alum workers and provided them with accommodation.

So what was life like for them?

Near Kirby Underdale, in the Yorkshire Wolds east of York, one of the most bizarre signposts in Yorkshire points the way to the village of Hanging Grimston. Drive down the hill and up the other side, and you find an identical signpost pointing back the way you have come. But you have not passed through a village, nor have you passed another junction. And the couple of farms along the road have different names. Where, then, is Hanging Grimston?

Drive slowly back down the lane, and you notice a few humps in an otherwise smooth grassy meadow. This is – or was – Hanging Grimston: a medieval village which numbered eighty souls over the age of fourteen in 1381, but which was 'cleared' in the 1580s, which is to say that its population was evicted to make way for sheep rearing, a more profitable use of land for the landowner. By 1619 the once thriving village was deserted, and today no buildings survive.[26] Why a signpost should point the way to a village that disappeared in the late sixteenth century is a mystery.

The fate of the inhabitants of Hanging Grimston is a stark reminder that working people in seventeenth-century England lacked even the most basic protection. England's first legislation to protect workers was the Factory Act of 1802; the alum industry pre-dated it by two centuries. Workers depended on the good will of their landlords and managers, which is why Hanging Grimston is pertinent to our story: from 1593, the landowner in question was Sir John Bourchier, the fourth patentee of the alum company. Sir John's treatment of the agricultural workers on his land at Hanging Grimston did not bode well for the workers in Yorkshire's alum industry.

Of course, the alum company claimed to have working people's best interests at heart. According to a document of 1609, the company paid not only thousands of its own workers a wage, but also sustained the coal miners at Harraton, the crews of the ships that transported the coal, urine and alum, and workers in the nation's cloth and leather industries. Moreover, asserted the company, it instructed its workers in religion and educated their children.[27] It paid William Ward eighty marks a year to minister to the alum workers in Guisborough, and Richard Leake, the minister at Lythe, to do the same for the workers at Sandsend and Asholm.[28] Not that this concern for their workers' spiritual needs prevented the company from requiring work on Sundays.[29]

In any case, the workers might have willingly exchanged their religious instruction for regular pay. For despite the company's pious words, an inquiry in 1612 found that the payment of wages was lamentably in arrears at the alum works, and that 'a great number of people [were] crying out for money that was due to them'. At Belman Bank alone, 26 workers lived virtually hand to mouth, not only because their daily pay of sixpence or eightpence was pitiful, but because they had 450 days of unpaid work between them. And the arrears of August 1612 were not paid till December.[30]

Five years later, it was the same story. Wages had not been paid for three or four months, and the workmen were described as 'poor snakes, tattered and naked, ready to starve for want of food and clothes', and unable to get credit for food and drink. Courageously, when they were offered one month's wages instead of the three or four that they were owed, they went on strike.[31]

In an age before there were trade unions to fight on the workers' behalf, it was in fact scandalously common for labourers to be paid many months in arrears. Even men employed to work for the king received their wages two years overdue. No wonder desperate workmen at the alum works threatened to resort to violence and sell the contents of the alum house to raise money to feed their families.[32] As Dr Edward Jordan, who ran the Slapewath alum works in 1614, wryly put it: 'The curse of the unpaid workman upon the works was more to be feared than that of the pope'.[33]

To make matters worse, labourers often were paid partly in kind: part of their 'pay' came, as yet another inquiry found in 1625, as 'malt, peas, corn and cheese ... instead of wages'.[34] Many were 'paid' with 'tickets' which could only be redeemed at the company food shop, where the company recouped its money by selling food at high prices. And those who lived in houses that belonged to the company might have part of their wage deducted for housing, again at rates set by the company.[35]

And life was quite tough enough for the alum workers, even without the issue of unpaid wages. It rained so much in the August of 1612 that floods washed away the timberwork at Slapewath, and the wreckage had to be hauled back up the hill. The winter of 1614/15 was so cold that the water froze in the pits and a thick covering snow made it impossible to mine the shale.[36] Moreover, the cliffs were notoriously prone to landslides. In spring 1682 the whole village of Runswick Bay slid into the sea, and at Kettleness the alum works along with the labourers' cottages were destroyed in a landslide in 1829.[37]

Accidents must have been common at the works too, though we have no early records. Walter White records that on one occasion 'the cliff took fire and burned for two years'.[38] A constant danger was being smothered by falling rock. This is what killed William Wales in September 1784, and Thomas Garbutt and Joseph Wardell in February 1824. Garbutt's widow and seven children were left with no financial support.[39] There were accidents at sea, too. A Tyneside collier ran aground on the rocks off Boulby in September 1827,[40] and a London brig called the *True Blue* went down with all hands in April 1845. When a Boulby alum worker stumbled across a decomposing human body eight weeks later, the corpse was assumed to be that of one of its crew.[41]

And then there was the matter of the unutterably disgusting smell. It was so powerful, so nauseous that the opening of an alum works in London in 1626, downstream from the Tower, in an effort to produce alum closer to the London dyers who were the main users, was met with howls of protest about the smell, and claims that the stench was so bad that cows in adjoining fields were unable to eat their grass. The President of the College of Physicians and six other doctors investigated and reported that the alum works 'endangered health'. So the London works closed in 1627.[42]

And as late as 1904, farmer Thomas Harrison took the Runcorn alum works to court for 'injury to his crops of corn and roots' from the 'noxious gases and vapours' emanating from the works which were processing alum imported from the largest deposits of alunite in the southern hemisphere, discovered at Bulahdelah in New South Wales in 1878. The jury found in farmer Harrison's favour and he was awarded £366 in compensation.[43]

Back in the seventeenth century, however, on the remote coasts of north Yorkshire, no court, no College of Physicians nor anyone else had a heart for the alum workers who laboured day in and day out with the smell which elsewhere was considered so harmful.

The production of alum in Yorkshire was undoubtedly a technological *tour de force*. But the financial system that initially underpinned it was a mess.

Having committed to produce 1,800 tons of alum a year, by 1608 the alum company's output was only 260 tons a year and the company was losing money hand over fist. But when the first of many inquiries looked into the alum industry, it reported – quite unrealistically – that production could be increased to 2,000 tonnes, which persuaded King James I to take the alum works into Crown ownership. And the king appointed Arthur Ingram, who had conducted the inquiry, to manage the works on the Crown's behalf.

In the new arrangement, Ingram contracted to produce 1,200 tonnes of alum in the first year, rising to 1,800 tonnes in every year after that.[44] But production actually decreased: only 575 tonnes were produced to March 1616, debts soared, and the £10,000 that the King had been persuaded to invest to repair buildings and equipment had disappeared without trace.

Ingram's contract imposed penalties on him if he failed to meet the targets he had agreed to, so at this point he should of course have been hauled before the Treasury to account for the shortfall. However, what counted in Stuart England was to have friends in high places, and, among several positions of rank that he occupied, Arthur Ingram was secretary of the powerful Council of the North, of whom Sheffield, one of the patentees, was president.[45] Even more importantly, Ingram had a patron in Thomas Howard, Earl of Suffolk, who, as Lord High Treasurer from 1614, had responsibility for the Crown's interest in the alum works. Suffolk, who lived a lavish lifestyle and spent a fortune on such grand palaces as Audley End House, was famously short of money and open to bribery: he was later dismissed for accepting £1900 from William Turner, an alum contractor, and £1,000 or £1,500 from Sir Francis Hildesley, a purveyor of iron to the alum company.[46] So maybe money passed hands, or maybe Suffolk was just returning a favour to Ingram: either way, Suffolk saw to it that Ingram escaped scot-free, and his penalty of £12,340 was quietly written off.[47]

And what happened next was more extraordinary still, for in 1617 Arthur Ingram was given the lease of the alum works for a further nineteen years. Once more his tenure was mired in scandal, with Ingram accused of embezzlement by Sir John Bourchier. In fact neither man could be trusted. One – Ingram – was infamous for acquiring large properties, paying half the purchase price and then refusing to pay the second half on the pretext that he had found some mistake in the deeds. As often as not the landowner backed down, rather than face ruin by going to court against a man with friends in high places. The other – Bourchier – was known for imposing outrageously high rents on his tenants, and was in and out of court on charges of shady dealing. 1625 found both Ingram and Bourchier under house arrest, probably following a punch-up. And in 1633 yet another patentee – Sir David Foulis – was also jailed, for non-payment of a fine.

It was in fact only after these gentlemen had been forced to withdraw from the scene that serious business practices were at last put into place. 1635 marked the first year that the magic production figure of 1800 tons of alum was actually reached.[48] The works then turned in a profit and continued to grow, despite the interruption of the Civil Wars in 1642-1646. One measure of the growth was that the local supply of wood was soon exhausted, requiring coal to be used instead in the burning of the clamps. Ten thousand tons of coal were brought in from Tyneside in 1700; by 1805 this figure had doubled to 20,000 tons.[49] And a second measure was that exports boomed. Being cheaper than Tolfa alum, and nearly as good quality, Yorkshire alum dislodged Tolfa from the lucrative Dutch market. And the alum works around Liège withered in the mid-18th century due to the stiff competition from Yorkshire alum.[50]

Many different dates have been put forward for the start of the industrial revolution in Britain. Was it the invention of the flying shuttle, patented in 1733? Or James Watt's and Matthew Boulton's improved steam engine of 1778? Or the building of the first iron bridge in the world in Ironbridge in 1781?

Yorkshire's alum industry pre-dates all of these key events by well over a century, yet clearly bore the hallmarks of an industrial revolution. New industrial processes were set in motion. New working populations were attracted and cheaply housed in company housing. Working hours were adjusted to make the most use of the industrial equipment. Long-distance transport routes were developed, wharves were built, tramways laid, industrial buildings put up. Capitalist practices, unbridled by law, created lots of new jobs and exploited the workers while they worked. Some owners became very rich, others were ruined. There was hard graft and malpractice. There were the first industrial strikes. And the alum industry even heralded the industrial north-south divide, with the workers labouring in the north, while the fruits of their labours, in the form of the colourful clothes and fabrics produced by the London dyers, were enjoyed mainly in the south.

Most dramatically, the alum industry presaged the wanton rape of the countryside that would be a characteristic of industrial-scale mining. Quarries were hewn out of the hillsides, great bites were taken out of the cliffs on the north Yorkshire coast, leaving scars that are clearly visible to this day. In places, the debris from the burnt shale is so thick on the ground that it has created a grey, moon-like landscape on which few plants will grow.

That, however, was all for the future.

Back in the first half of the seventeenth century, the alum works prospered, and an increasing number of ships plied the east coast, bringing building stone to the Tyne and Wear, coal to Whitby, alum to London and urine back to Whitby.

And Captain Cook was not the only one to make wider use of the seafaring skills acquired in this trade. In 1631 we find Captain Luke Foxe, who we met modestly 'taking the piss' in his ship the *Allome Amye*, setting off from London aboard a ship called the *Charles*, bound for the coast of Canada where he would spend July, August and September exploring Hudson Bay.[51] Had he discovered a Northwest Passage from the Atlantic to the Pacific Ocean, and had the great land mass of Australia not existed, the statue on Whitby's west bank would have been of Luke Foxe instead of James Cook. No matter. Both explorers had cut their teeth in the alum trade.

So, thinking about it, I will take Whitby's statue of Captain Cook to honour the alum trade after all, even if the word 'alum' does not feature on it.

14

Alum, India and Global Trade

Far back in the mists of time, so the story goes, a man called Qin Fu had to flee his native Sichuan province, in the west of China, along with his wife and children. The family crossed the length of China and finally found shelter in a cave in the Jilong Mountains, close to the southeast coast. Here they put down their things, made a fire, collected a few stones on which to balance their cooking pot, and cooked themselves a meal.

A few days later, the family were surprised to see that the stones they had collected were beginning to disintegrate. Then, to their even greater astonishment, the stones began sprouting small crystals which, when they licked them, they found to be sour. And when Qin Fu threw the crystals into a puddle, not only did they dissolve but, as the family watched entranced, they caused the muddy water to turn clear, as if by magic.

Later, Qin Fu's son had a stomach ache and a bad bout of sunstroke. His parents gave him a bowl of water into which a few of the 'magic' crystals had been dissolved, and again the crystals appeared to have magic powers, for the child soon felt better again.

The crystals were of course alum crystals. Qin Fu had stumbled upon Wenzhou, the largest alunite reserves in the world. After six hundred years of commercial exploitation they still contain 60% of the world's known alunite.[1]

At least seven different works exploited different veins of alunite at Wenzhou at different periods. But it was always a massive enterprise, with some of the mines reaching 1100 metres into the mountainside.[2] The vast site also included the alum processing works and entire villages for the alum workers. Very sensibly, knowing what we do about the evil smell of the alum works, the villages were situated upwind of the works.[3]

The processes used at Wenzhou were broadly the ones we are familiar with. The ore was first calcined (burned) and then dissolved to create a liquor from which the alum was crystallised. Two key differences were that the calcination took place not in open-air clamps, but in large metal kilns, and the crystallisation took place in circular ponds, rather than in enclosed roaching stacks. Rows of gaunt-looking kilns still tower today over the site which, together with the alum workers' village of Fudewan, is now a conservation area.

Alum was produced at Wenzhou in such large quantities that the works not only supplied dyers in China, but exported alum to countries across Southeast Asia. Like the Genoese, the Chinese merchants used alum as ballast for their shipping and then sold it as a profitable commodity at the port of destination.[4]

Which is how, from at least the sixteenth century, Chinese alum nurtured an industry several thousand miles west, on the east coast of India.

Cotton cloth was manufactured in India from as far back as 5500 BCE but we do not know what sorts of mordants were originally used. From the ninth century C. E. at the latest however, Indian dyers in southeast India were importing their alum from China,[5] and what's more, they had mastered the art of using alum to fix the colours in patterns which they printed on cotton with wooden blocks. Thus, when the first European merchants arrived on the east coast of India, they found a thriving trade in colourful printed cotton fabrics, with ports such as Masulipatam and Pulicat, north of Chennai, exporting printed cloth to Indonesia and other countries across the Indian Ocean. Indeed, Indian weavers were already producing patterns to overseas specifications, as stipulated for example by the court in Siam (Thailand).[6]

Now, although the focus in our stories so far has been on woollen cloth, Europeans knew – or thought they knew – what printed cotton looked like. Printed cotton fabric from Aleppo was traded in Marseille. Armenian weavers there produced imitations of cotton imported from

Egypt. However, being lighter to wear than woollens, printed cotton fabrics were soon popular across Europe, and by the late 1600s mills in England and Holland were producing them too. The first opened in Holland in 1678.[7] Like in India, carved wooden blocks were used to apply the colour, and there was no lack of dyes, as Europe now had access to those brought by ship from the East Indies as well as those traded along the Silk Road. But the mordants were a problem. For alum to work on cotton it had to be thickened with resin, which had the effect of darkening the colour of the dyes. And even then the colours were not always fast.

Which is why the European merchants were astounded by the cotton fabrics they found on the Coromandel coast of southeast India. They were of a quality that outclassed anything that they had seen before.

The cotton fabric itself was lighter in weight than the European product, because the Indian weavers used age-old knowledge to treat the fabric, involving up to seventeen different steps. They repeatedly beat the fabric on a stone, boiled it, soaked it in sheep dung by night, laid it out in the bright sunlight by day, then boiled, washed and soaked it again, first in a solution of tannin-rich fruit, then in buffalo milk. They wrung it in several directions, exposed it to sunlight, and then rubbed it firmly to flatten and smooth the surface.[8]

What took the Europeans' breath away even more than the lightness of the fabric, however, was the miraculous brilliance of the colours. They shone with a power and luminosity that made anything produced in Europe look shabby, and the colours were fast. And this was all down to the Indian dyers' expert use of alum.

The Indian dyers applied patterns with wooden blocks and used resin and other gums to thicken the mordant, as did the Europeans. But the special treatment to which the cotton cloth had been subjected allowed the mordant – fermented iron for the black outlines, alum for the very fashionable red outlines – to penetrate deeper into the fabric. The dyers then laid the fabric in a bath of madder dye to allow the background to revert to a cleaner white, against which the colours shone more radiantly.[9] And crucially, knowing that the resin had the effect of darkening the colour of the dyes, the Indian dyers had perfected a way of washing out the resin after the colour had been applied. This vital step gave the Indian colours an unparalleled vividness compared with European cottons, in which the resin 'saddened' the colour.[10]

As a result, Indian fabrics were traded eastwards across the Indian Ocean and westwards to the east coast of Africa, where they were sold inland as far as the Congo.[11] And once the merchants of the English, Dutch and French East India Companies loaded them onto their ships, and then displayed them in London, Amsterdam and the Breton port of Lorient (the seat of the French East India Company), the Indian fabrics hit Europe by storm.

Up to ten cargoes a year arrived just in Lorient,[12] and more than two million pieces were being imported annually into Europe by the late 17th century. Soon, to everyone's surprise, printed Indian fabrics had become the principal commodity traded between Asia and Europe, eclipsing even Oriental spices and Chinese ceramics.[13] By 1684, a staggering 73% of the trade of the English East India Company was in Indian cotton cloth.[14]

Known as *indiennes* in France and *calico* or *chintz*[15] in England, the fabrics were initially used in furnishings, as curtain material, bed hangings and in upholstery, replacing the heavy embroidered brocades that had been in fashion previously.[16] But when the new fabrics were then worn by working people, who made clothes from the cast-offs given them by the families they worked for, the rich suddenly saw their potential for beautiful clothing, and they adopted the style too. Thus everybody ended up wearing the light Indian fabric clothing – and Indian *chintz* had triumphed over the long tradition of using colour to divide human beings by class or gender, creating, in the words of Sarah Fee, 'the first mass fashion'.[17]

Not that everyone approved, of course. The curmudgeonly took umbrage, as they do, with Daniel Defoe, for example, muttering disparagingly about 'persons of quality dressed in Indian carpets'.[18]

And of course, the owners of European textile mills were mightily displeased, as were landed aristocrats with large flocks of sheep: they both saw their business model collapse as people fell head over heels for the imported cotton product.[19] These were powerful lobbies, and they soon put pressure on governments to come to their rescue. The French government at the court in Versailles banned imports of *indiennes* in

1686. In England, after five thousand angry weavers descended on Parliament, the Calico Act of 1700 restricted imports of Indian chintz into the country. The rules were tightened further in 1721.[20] In Prussia, King Frederick William ordered that all Indian cotton fabrics in his kingdom should be destroyed within eight months.[21] Spain, Venice and the Ottoman Empire all took similar action.[22] But the tide could not be turned back. The fashion had caught on too strongly. The bans simply led to floods of illegal imports, smuggled in from countries where no ban was in force: from Switzerland into France, and from Holland into England.[23] And, in any case, the bans were routinely flouted at court, by those important enough to be able to do so.

One of the effects of the bans was to encourage the importers of Indian fabrics to find a new outlet for their wares, and they found an easy market among fellow merchants who were involved in a different trade altogether: the heinous cross-Atlantic slave trade. The bans on importing *chintz* and *indiennes* did not apply if the cloth was imported and immediately re-exported, so all the French East Indies Company merchants, for example, had to do was unload their fabrics in Lorient and then ship them down the coast to the slavers operating from Nantes for export to Africa. And this soon became the pattern in all countries involved in the slave trade. The slavers carried the Indian fabrics to the West African coast, where they were exchanged for the slaves who were then transported to the Americas.

The alum from far away Wenzhou, therefore, stimulated a cotton industry in India whose products were sold across east Asia and took Europe by storm, triggering government action which diverted the trade to Africa where it became embroiled in the trans-Atlantic slave trade. Alum was thus the essential link in one of the first truly global commercial networks that connected China to India, India to the East Indies and east and central Africa, India to Europe, and Europe to west Africa and the Americas.[24]

15

Alum, Two Women and a Monopoly

Our many stories of alum, for all that they have ranged across the centuries, across cultures and across the globe, have had one constant. The chief protagonists have been relentlessly male.

But this changes in the eighteenth century. And most dramatically in Skane, now the southernmost province of Sweden but for centuries part of the Kingdom of Denmark, where a woman is still feted today as one of the most powerful entrepreneurs ever to have run a major industry.

The story begins with a rich man called Jochum Beck, who was sleeping one night in 1635 on his estate of Andrarum, in Skane, when he dreamt he was lying on treasure. The next morning Jochum was shown a piece of alum, and he needed no more convincing. Jochum Beck bought up land around Andrarum, raised a loan in Holland, obtained a licence from his friend the King of Denmark, imported experienced German alum workers and launched Denmark's (and Sweden's) first alum works.

Then war threw a spanner in the works. Denmark clashed with Sweden in 1658, both armies plundered the works (reputedly melting down the alum pans to make bullets), and Skane passed under Swedish control. When the Dutch banks called in their loans, Jochum Beck was unable to pay up and he died bankrupt in 1682.

Which is where a widowed woman by the name of Christina Piper (pronounced *Pee-per*) comes into the story. She took over the Andrarum alum works in 1725, and she came with a formidable reputation.

Born in 1673 into one of the richest merchant families in Stockholm, Christina had been married off aged 17 to Carl Piper, who was fully 26 years older than she was, but who had excellent connections at the Swedish royal court. And Christina made the most of it. It soon became known that if you desired a word in the king's ear, you could pay Christina a social call and leave her an expensive – what shall we call it? – gift.

Christina would get Carl to arrange an audience with the king, while Christina invested the, err, gift in a growing portfolio of property.[1]

But this comfortable arrangement came to an abrupt end. In 1700 King Charles XII of Sweden went to war with Russia with disastrous results: at the Battle of Poltava in 1709 Sweden lost its army, Carl Piper lost his freedom and Christina lost her husband. Carl was taken prisoner and he died in gaol in Russia in 1716.

How did Christina feel about this sudden change in her life? Many will have pitied a widow with three daughters and a son to raise without the support of a man. But Christina took control of her life as she had never done before, and in her 50,000 acres in Skane she made the alum works in Andrarum the centre of a mini-empire run by herself.

Christina expanded the workforce to 900 employees, and had a school built, as well as a pharmacy, a hospital and a retirement home. Together with Andrarum's own court and prison. And a flogging post.[2] Christina paid her workers with her own coins, pressed in her own mint, embossed with her own initials. This was the only currency accepted in her company's stores and canteen. The workers may not have liked it, but were in no position to argue, with the grain harvests failing in 1740-1742.[3] It was an age of notoriously harsh winters, but by royal decree, all the wood grown within twenty miles of Andrarum had to be delivered to the alum works, at a price set by the works owner. The peasants were not allowed to keep any for themselves.

By the time Christina Piper died in 1752, aged 79, the alum works at Andrarum had, under her management, become one of the largest industrial sites in Scandinavia. All told, some 3,000 people lived and worked in and around the alum works, an extraordinarily high figure when you consider that the regional centre, Malmö, had a population of about 5,000 at the time.

And Christina's control over her estates outlived her mortal life. Not trusting her – male – descendants to be as financially savvy as she had been, she entailed the Andrarum estate and alum works in a legal framework that ensured it could not be sold or sub-divided, in perpetuity. And the estate is intact to this day.

In faraway Yorkshire, meanwhile, a less powerful but equally resilient woman also became involved in the alum industry.

At about the time that Christina Piper died, Judith Routh married George Baker, a man with an interest in horses. And as Judith's father was a successful breeder of racehorses, Judith knew exactly what that meant. She was no novice to the intimate bond between horses and gambling.

Which was maybe why, in the Baker household, it was Judith who kept the family accounts while George was out training his hounds for hunting.[4] It was she who paid the tradespeople when they called, and who oversaw the craftsmen remodelling Elemore House, George's family home in County Durham, in the fashionable Palladian style. And it was Judith who managed the sale of George's properties in Northumberland. Judith had a good head for figures, and was indeed as tight-fisted as the next man when it came to paying her bills: her glovemaker in London complained in November 1769 that he had not been paid for sixty pairs of gloves he had supplied since April of the previous year. On the other hand, Judith appreciated loyalty, and preferred to return to trusted suppliers rather than to change to possibly cheaper ones.[5]

Nor was it as uncommon as we may think for educated women to take charge of their family's finances, even if less publicly than Christina Piper in Sweden. Christine Wiskin has noted that many women of Judith's rank 'quietly got on with the highly rational business of keeping their families solvent'.[6] What was unusual in Judith's case was that she also managed George's business – and that business was alum.

By the mid-eighteenth century, the Crown monopoly on alum in England had long gone,[7] and the Yorkshire mines were largely in the hands of such wealthy landowners as the Mulgraves at Sandsend, the Dundas's at Kettleness and Loftus, and the Cholmleys at Saltwick, near Whitby. As landed gentry,[8] the Bakers were, socially speaking, a cut below these powerful families, but George had inherited the principal share in the Boulby mine, a smaller alum mine which had first opened in the 1650s and whose workers lived, for the most part, in nearby Staithes. It was separated from Lord Dundas's Lofthouse alum works to the north by a narrow neck of rock.[9]

When her husband became deputy Lieutenant for County Durham in 1754, Judith took over as managing partner of the Boulby works, working closely with Ralph Jackson, a local man who owned a quarter share in the mine. In a successful business partnership that lasted twenty-eight years, Ralph saw to the day-to-day business at the mine, while Judith was responsible for the company accounts and forward planning.[10]

Judith and Ralph saw at once that their mine was crying out for investment. But investing in an alum mine in the eighteenth century was a perilous business. You had to borrow money but you had no control over the many factors that could determine whether your mine would be profitable or not, so you easily end up in debt.

And indeed, no sooner had Judith borrowed money and mortgaged some of her assets to invest in the Boulby mine than the Seven Years War broke out in 1756, and the supply of workers drained away. Some maybe volunteered for service, but when the Royal Navy needed seamen, it pressured or simply kidnapped young men from towns and villages along the coast, leaving families deprived of their breadwinner, and employers of their workforce. Very likely lured aboard a navy ship during an evening out in Staithes, a number of Boulby alum workers found themselves among the nearly 75,000 men (!) press-ganged into the Navy in the course of the war.[11]

Then, in June 1760, French privateers made off with the Boulby alum company's sloop *Darling*, taking not only its cargo but also its captain Anthony Jefferson and his family.[12] Such were the realities of running what might be called a 'modern' business in the eighteenth century.

And yet the Yorkshire alum trade persisted, with a *Description of England and Wales* written in 1770 noting that twenty to thirty ships took Yorkshire alum to the Baltic, nine or ten boats to Holland, and five or six to the Mediterranean.[13]

⁂

Now, imagine for a moment that you are Judith Baker. The year is 1771, the war is over, and you have a decision to make.

The town is abuzz with talk of a new scheme being put to Yorkshire alum mine owners such as yourself. In a nutshell, the deal is this: if you commit to selling all your alum to just one company, the scheme will in return pay you a guaranteed income of £14 per ton.

The question is: should you join the scheme or not?

On the negative side, you are maybe reluctant to put all your eggs in one basket. The scheme demands *all* of your alum, so you would have to abandon your loyal London customers of many years. Loyalty, remember, is a quality you set much store by. And of course you want to know who is behind the scheme. In this respect the news is reassuring. While the scheme operates under a newly formed company called Ambrose Lynch Gilbert, Brown & Co., it transpires that the man behind it is none other than Sir George Colebrook, chairman of the East India Company, the largest and most successful company in the Western world.[14]

Being Judith, you do your sums. Since investing in your mine, you and Ralph have been selling your alum for around £11 a ton, sometimes £12, so the new scheme would pay more. And it would take between 290 and 330 tons of alum per year, more than you have ever produced before, because the low sale price never warranted an increase in production. You crunch the figures, and work out that even if you only produced the lower figure 290 tons, you would make £870 more in your first year. If your production reached 330 tons, your profit would be even higher. The investments you made back in 1756 would finally pay off.[15]

So, if you were in Judith's shoes, what would you do?

Judith and Ralph signed up for the scheme.[16] So did all the other alum mine owners in Yorkshire. Only Lord Dundas refused.

⁂

Colebrook's scheme was the latest incarnation of a well-known commercial strategy: corner the market in a key commodity, force its price to rise, and make a killing when you sell. But it did not have a good track record in Yorkshire.

Early in the eighteenth century Lord Mulgrave (now graced with the title of Duke of Buckingham) had tried to drive up the price of alum by eliminating the competition, paying the Peak works in Ravenscar £430, the Carleton works £400, the Saltwick mine £220, and other mines too, to close down for 21 years. In the short term it paid off, but then two things happened that Lord Mulgrave had not foreseen. One was that some dyers turned their back on his dearer alum and bought cheaper alum from abroad instead. And the second was that disused alum mines in Yorkshire re-opened, because the higher price of alum made them economically viable once more. The upshot was that Lord Mulgrave in fact ended up with more rivals than he had had before, and as total production rose, the price of alum slumped back to £10 per ton, and the scheme collapsed in 1736.[17]

Undeterred, the owners of Yorkshire's four surviving alum works then clubbed together in what we would call a cartel, agreeing to limit the overall production of Yorkshire alum to 1,500 tonnes a year, with each works allocated a fixed quota. Again they succeeded for a while, seeing the price rise to £12 in 1740, and £14 per ton in 1746; but again dyers turned to foreign competition, again disused Yorkshire mines came back into use, and again the benefit was short-lived.[18]

So what made Sir George Colebrook's scheme different from these failed attempts? Its sheer scale. Indeed, with true eurocentric blinkers, his scheme is often described as a 'world' monopoly of alum, which it patently was not – the alum producers of Wenzhou had probably never heard of it. Nevertheless, Colebrook's plan was to buy out all British production and the entire European market as well. As in the previous schemes, he would buy while alum was cheap, drive up the sale price, and then sell it at a huge profit. But this time dyers would not be able to turn to foreign suppliers, because Colebrook would have bought them out first. And the price rise would be so steep and sudden that disused mines would simply not have time to re-open and start a rival production.

The plan required huge amounts of money, of course, to buy up such vast reserves of alum. But this was no problem for Sir George Colebrook. As Chairman of the East India Company, he was one of the most solidly credit-worthy men in the world, so he was able to raise the necessary credit for his alum scheme, chiefly from merchants in London and Amsterdam. Indeed, so successful was he that there were calls for a new institution to formalise the means of raising capital. It was set up in London the following year. Yes, we have alum to thank for the creation of the London Stock Exchange.[19]

Having got all but one of the Yorkshire mines on board, as well as a smaller alum operation in Lancashire, Colebrook cast his net abroad. In January 1772 the mines around Liège contracted to sell him their production of 576 tons[20] of alum per year, and in February and April he struck a similar deal with Charles Augustus Piper, Christina Piper's son, for all the alum from the *Garphytta* and *Andrarum* works. The *Kaflas* works in Sweden joined in too. In May he negotiated with the Jewish firm Solomon and Abraham Alasdrak to buy up all the alum traded at the port of Izmir (Smyrna), the principal Ottoman port for exporting alum since Phocaea had closed down. The alum will have been brought to Izmir by camel from Gediz (formerly Kütahya) and Sebinkarahissar (formerly Koloneia).[21]

Even the famous Tolfa mines succumbed to Colebrook's grip. They had been run since 1715, and possibly with the use of Muslim slave labour,[22] by the Lepri family, who had standing contracts with the Vatican and the French company Gachon & Cie.[23] But Marchese Giuseppe Lepri was happy enough to sign away all the remaining Tolfa production, estimated at 6,000 cantara a year, to Colebrook's scheme in April 1772. Colebrook then approached Gachon & Cie directly and snapped up the 6,000 cantara of alum that Lepri was contracted to sell them. Colebrook had thus secured the entire Tolfa production, save only the supplies to the Vatican. He now had a virtual monopoly on European alum production.[24]

In order to sell his alum, Colebrook set up two sales hubs, one in London, and one in Amsterdam, where his bankers were the well-known Clifford and Sons. The hubs commissioned agents around Europe, from Rotterdam to Riga, from St Petersburg to Smyrna, and from Copenhagen to Cadiz. In March 1772, Clifford & Son made loans to the tune of £20,000, guaranteed by Colebrook, and the scheme finally swung into action.

And the price of alum soared. The *London Evening Post* noted a steep rise as early as April 1772, and the newspaper confidently predicted that alum might later in the year hit the heady heights of 30 guineas (£31 10s) per tonne. As the company had acquired 6,894 tonnes of alum, worth £147,000 at the lower cost price, it looked set to make a jaw-droppingly successful profit. Successful for the speculators, that is. The investors' joy at the skyrocketing price of alum was accompanied by the manufacturers' dismay that the scheme had ruined 'many worthy tradesmen'.[25]

The investors' joy was short-lived, however, for it soon transpired that the scheme was beset with serious problems.

One was that Colebrook's claim to have cornered the market in Europe turned out to be wishful thinking. In fact one-third of European production was still out of his hands, and these producers could steal his customers by undercutting his high prices. Moreover, with the cost of alum so high, customers were postponing their orders, so while the price of alum was impressive on paper, very little had actually been sold: only 29 tonnes were sold in England between April and October 1772, for example. And, with sales so sluggish, more time had elapsed than had been forecast, giving disused alum mines, after all, time to re-open and to sell their alum at reduced rates.[26]

All the above were commercial consequences that a well-funded operation should have been able to weather. But well-funded it was not. On the contrary, it soon became apparent that the scheme was built on wild speculation. Colebrook had borrowed massive sums of money (about £166,000) by December 1772, and funds were required to service this debt. But Colebrook had hoped to raise these funds from winnings on bets that alum would be selling at £30 per ton within six months, and that 500 tons of alum would have been sold by August. The second bet alone, had it been successful, would have netted a cool £60,000 which he could have used to pay his creditors. But it wasn't, it didn't and he couldn't. The targets were missed, the winnings on the bets never materialised, a big black hole appeared in the company's accounts, and the scheme began to run out of money.[27]

Rumours now spread that the scheme was unsound. As early as July 1772 Colebrook's partner Ambrose Gilbert was unable to raise a new loan in Amsterdam and Paris. By September Colebrook's agents were encouraged to sell their alum at a reduced price, which of course signalled to all and sundry that his alum scheme was in trouble. And at this point, the scheme became embroiled in an even wider banking crisis that had been gathering momentum during the year.

Banks had been unsettled when a speculative banker by the name of Alexander Fordyce fled the country in June 1772, and a number of Scottish banks collapsed when his fraudulent activities subsequently came to light. Then, in September, when the East India Company announced that it would issue no dividend, its share price plunged, wiping out the savings of thousands of investors who had invested in it. This caused further alarm. The Bank of England refused to renew a rolling loan of £400,000, and in October the East India Company was obliged to call in funds from its other investors. This in turn caused panic in Amsterdam, where many of these investors were based, and in the winter of 1772-3, the bank of Clifford & Sons went bankrupt.[28]

The alum scheme, and the East India Company ... The two big institutions that found themselves in deepest trouble were both headed by Sir George Colebrook, and that was no coincidence. For just as he had taken bets on the success of the alum scheme, he had also speculated wildly on the stock of the East India Company. By the end of 1772, Colebrook's losses in the alum scheme alone stood at over £100,000, and, amazingly, it then transpired that he had already accrued debts of £190,000 while simultaneously speculating on hemp. Colebrook's friends in high places managed to keep him afloat till 1773, but then his business empire came crashing down. Colebrook was forced to sell his family home and a residence in Soho Square, and flee with his family to Boulogne, leaving a trail of creditors back home, some bankrupted, others in deep financial trouble.[29]

One of those left to pick up the pieces was Judith Baker.

Alarm bells had first rung in her mind when, out of the blue, Ambrose Gilbert had issued a writ against her for allegedly having broken contract with him. Judith knew very well that she had not. But the writ had quickly become an irrelevant side-show as the full scale of the disaster unravelled and hit home. Dragged under by the collapse of the Colebrook monopoly, her customers had become unable to pay for the alum they had been supplied with, and Judith's rosy profit forecasts withered on the vine. The firm of Sam Schutz, who had taken 40% of her alum under the scheme and owed her £441, was failing. Hawkins and Byrne, who owed £236, were already bankrupt. William Cutlove (owed £52) was described as 'at present poor', James Rea (owed £6) was 'not worth a shilling', and one named simply as Nugal (owed £11) had 'absconded' and 'fled the country'. Judith had to count herself lucky that she had agreed a payment of ten shillings (half a pound) in lieu of Thomas Morriss's debt of £22.[30] The sales side of the business had to be rebuilt from scratch.

And yet, while creditors of Sir George Colebrook forced alum works such as those in Guisborough to close, the Boulby alum works survived.[31] This says much for Judith, for her partnership with Ralph Jackson which remained intact despite the upheavals, and for the steady hand of Boulby's new mine agent, George Dodds, who had been appointed in July 1772, just before the storm broke over his head. Above all, it highlights the loyalty of those who had invested in Judith's business, and who could have taken fright and withdrawn their money, but instead trusted her to rebuild the business. 40% of the investors were women, and it would be fascinating to know whether the fact that Boulby's senior partner was female played a part in their decision.

And indeed Judith and Ralph Jackson pulled the business round, despite further blows such as debts of £6,000 left by Judith's husband George when he died in 1774, and the discovery that Thomas Core, the company's agent in London, had been building up a network of contacts for his own benefit, rather than the company's. By the time Judith retired to Tynemouth in 1784 the company was on a firm footing once more, [32] and when she died in 1810, Boulby was one of the six alum works that had survived in Yorkshire, producing 450 tons a year towards the total production of 2840 tons.[33]

In Sweden, Christina Piper is celebrated as a larger-than-life figure.

In England, Sir John Colebrook, who ruined so many lives and businesses in England and abroad, negotiated his return from Boulogne to England, where he retained his knighthood and re-entered high society in 1786 as one of 'the artistic and literary elite' in Bath.[34] His name brings up nearly 10,000 results on Google.

In contrast, Judith Baker remains virtually unknown. Yet she was a remarkable woman. She ran the Boulby alum works at a time when, as Susan Beaumont notes, 'there were no precedents for managing such enterprises'.[35] She provided workmen in Staithes and the area around Boulby with regular employment, and rescued her business from the damage caused by Colebrook's attempted alum monopoly. And she did this while bringing up two children, dealing with her husband's debts, and playing the role of the campaigning hostess, giving dinners in order to persuade those entitled to vote to rally to the reform wing of the pre-1832 Reform Bill parliament.[36]

We owe a debt of gratitude to those like Susan Beaumont and Christine Wiskin for having rescued this strong, courageous woman from undeserved obscurity.

16

Alum, Bread and Dinosaurs

Kettleness, 1857

Was there something funny in the air in shale quarries? We labourers certainly began to think so. Or maybe our overseers had a strange sense of humour. Because every so often they would order us to stop working and step back just in case we were hacking away at – wait for it! – a crocodile. A crocodile! Who in their right mind would be looking for crocodiles in Yorkshire?

Snakes, yes. Snakestones were common enough in the alum shale we were digging every day, and we all knew St Hilda had made them by turning snakes into stone when she became abbess of Whitby abbey.[1] But crocodiles! It was always a false alarm, of course. And not an unwelcome one, to be honest. No skin off our nose, having a bit of a break, what with us being paid by the day, not by the number of barrow loads. And it gave us a good laugh.

And then one day, there we were, leaning on our spades and pickaxes, smirking behind our overseer's back as he went down on all fours – when he suddenly went crazy with excitement. This time, he said, we had stumbled on something special. And when he put us back to work, it was at a different part of the quarry. He didn't want his find tampered with. His find, you'll notice. Our find had become his pretty quickly.

The next day some well-dressed folk from Whitby arrived, and, fine clothes and all, they also scrabbled around on their hands and knees where we had been working. Soon we saw them removing and labelling little pieces of rock, which they put in a trunk. And when they finally left, taking their trunk with them, our overseer was in an uncommonly good mood. 'Drinks on me this evening, lads,' he said. 'The gentlemen from Whitby have been very generous.'

And that is how we alum workers found one of the oldest and largest creatures you'll probably ever see in your life: the 180 million-year-old, eight-metre-long ichthyosaur, and pride of the Yorkshire Museum in York.

It was bound to happen. Shale, which is basically compressed mud, is typically rich in fossils, so with teams of alum workers picking and scraping at the shale along thirty miles of the Yorkshire coast, it was only a matter of time before the fossilised remains of creatures we loosely lump together as dinosaurs would become a somewhat surreal by-product of the Yorkshire alum industry.

And it is no coincidence that the creatures started being recorded in the mid-eighteenth century, and not before. By the 1750s a number of people in Whitby had become wealthy enough on the back of the alum industry to have the leisure to pursue their interests in the natural world, the means to pay for the fossils to be removed and the contacts with universities and museums eager to acquire the specimens for their institution.[2] Thus when in 1758 William Chapman found fossilised bones in what he called the 'alum-rock', they did not disappear into garden walls or into some private collection, as might have happened before. Instead, Chapman's find was classified, preserved and published in journals, and the fossilised bones of his marine crocodile ended up in the Natural History Museum in London.[3]

Whitby was now better connected to the rest of the country because, thanks to pressure from the *nouveaux riches* landlubbers, the land route across the moors was converted in 1764 from a 'rugged, miry, and uneven' track on which it was 'dangerous to approach the town on horseback ... still more so with a loaded carriage' into a turnpike road that allowed excited fossil hunters and others to 'pass, without danger or fear, over those deserts, which no stranger before that time ever durst venture to cross without a guide'.[4]

And the fossil hunters were well rewarded. A second crocodile came to light in 1791, parts of marine reptiles known as ichthyosaurs were found, 'imbedded', as the Reverend George Young recorded, 'in the alum-rock' in 1819 and 1821, and yet another marine reptile, a plesiosaur, followed in 1822. A local carpenter, Mr Brown Marshall, even made a bit of a name for himself as kind of local expert to be consulted each time an important find was made, and in 1824 Marshall himself recognised the snout of an ancient crocodile poking its nose out of the cliff at the Saltwick alum quarries. Known as a teleosaurus, it is now a star exhibit at the Whitby Museum.[5]

By 1828 the Yorkshire alum shales had released fragments of no less than seven crocodiles, forty ichthyosaurs and four plesiosaurs. Another plesiosaur was found in 1841, and the eight-metre-long ichthyosaur discovered by the alum workers in Kettleness, now in the Yorkshire Museum, was found in 1857. And indeed finds continue to be made to this day. More recently an ichthyosaurus was discovered by fossil collector Liam Langley at Runswick Bay in 2011.[6]

In the early nineteenth century the outlook for the Yorkshire alum industry appeared rosy. There were seven works in operation, in which more than five hundred workers produced 3000 tons of alum a year.[7] And the alum ports were thick with ships bringing coal from Newcastle and Sunderland.

Dyers and tanners in London were still by far the principal customers for the Yorkshire alum, but the industry had also found a new outlet in England's growing paper-making industry. The Chinese, as we saw in Chapter 3, had used alum to size their best quality Xuan paper back in the eighth century, and the Muslim paper makers who brought the knowledge to Europe also sized their paper with alum, dissolved in either starch or egg white.[8] Paper-making reached England in 1495, and English paper was typically sized 'in the German manner' with a mixture of alum and gelatine. Alum gave the paper more 'body' and made it crisper, while the gelatine protected the fibres in the paper from the alum's acidity. Just one paper mill, Whatman's Turkey Mill in Maidstone, Kent, was buying almost three tons of alum a year towards the end of the eighteenth century.[9]

From the 1830s, paper makers began to save money by sizing their paper with alum mixed with rosin instead of gelatine. This resulted in the brittle, discoloured paper that we now typically associate with old newspapers and documents, because the rosin did not protect the paper's fibres as well as gelatine. Whatman's wealthier customers, however, continued to use the top-grade paper, which is why Queen Victoria's letters, J.M.W. Turner's sketches, Napoleon's will, George Washington's documents and the Soviet Union's 1930s five-year plans have remained in pristine condition. They all used paper sized with alum and gelatine.[10]

Alum was also used by the plaiters in and around Aylesbury to colour the straw which they prepared for Luton's hat industry.[11] And Walter White knew of other uses too. When he visited the Yorkshire alum works in 1861, he noted that the alum produced there was used by 'dyers, leather-dressers, druggists [and] tallow-chandlers' and by 'bakers ... and other crafty traders'. Yes, alum was regularly being used to adulterate bread. So widespread had this scandalous abuse become, in fact, that in the nineteenth-century alum came to be known as 'baker's stuff'. And when Karl Marx published *Das Kapital* in 1867, his references to alum were not in the context of cloth, or leather, or paper, but solely in connection with adulterated bread.[12]

The adulteration of food was not a new problem, of course, nor specifically a British one. The unscrupulous miller is a well-known figure from Chaucer onwards, and in the late nineteenth-century salt merchants as far away as China 'enhanced their profit' by adulterating salt with dirt, gypsum, and alum.[13]

But the issue of adulterated bread took on particular importance in Britain in the nineteenth century because this was the age in which society became urbanised and came to rely on factory-made bread. In 1801, 20% of British people lived in towns; by 1901, only 20% did not. One consequence of this radical change was that whereas people had previously eaten food they had produced or made themselves, more and more people now relied on food that they bought. If bakers tampered with the bread they baked, the effects were therefore more serious than before.[14]

And tamper with it they did. Flouting a law of 1757 which banned the addition of additives, bakers routinely added an array of unsavoury ingredients: white clay, chalk, bone dust, bone ashes, plaster, soap, sawdust, dried horse manure and 'similarly pleasant, nutritious and healthy ingredients', as Karl Marx sarcastically put it in *Das Kapital*.

Above all, however, they added alum. And it's easy to see why. Like the other additives, alum helped to make the flour go further and added to the weight of the bread (and bread was sold by weight as from 1822). Alum, however, had the added advantage of making the bread look white, even if all sorts of doubtful ingredients had been added beforehand. And this was important because white bread was in vogue, with everyone wanting to imitate the bread eaten by the rich. Not surprisingly, therefore, when the physician Arthur Hassall analysed seventy-four loaves in 1857, he found that each and every one of them had had alum added. There was alum even in loaves made by bakers who had used none, because millers were also in on the act. Of eight samples of flour tested by Hassall, four contained alum.[15]

It took a German chemist living in London, Friedrich Accum, to publicise the scandal. Accum had actually come to London to establish the city's first gasworks, but in 1820 he published *A Treatise on Adulterations of Food and Culinary Poisons* in which he highlighted the role of alum in bread: 'To produce the degree of whiteness rendered indispensable by the caprice of the consumers of London, it [was] necessary ... that the dough should be bleached, and no substance [for] ... this purpose [was] better than alum.'[16] Accum daringly listed the names of merchants who had been prosecuted for adulterating food, but depressingly, the consequence was not that food standards improved, but that Accum was hounded from the country in 1821 on trumped-up charges of embezzlement.

Neither alum nor any other of the additives were in any shape or form wholesome, of course, and in 1855 a Committee of the House of Commons was told that the bread sold to the poor contained only a quarter of the nutrients it should.[17] Nor did the additives even make bread cheap. On the contrary, in 1857, when a labourer might earn twelve shillings a week, a loaf of bread cost a shilling – or one-twelfth of his weekly wage.[18]

Accum described this for what it was – theft. Why, he asked, should 'the man who robs a fellow subject of a few shillings in the highway ... be sentenced to death, while he who distributes a slow poison to a whole community should escape being punished'?[19] Richer households employed their own cooks, so it was the poor who were the chief victims of the crime. Crowded into the fast-growing towns, they no longer had the space, the time nor, very often, the ovens to make their own bread, so were forced to buy. The Cratchit family, in Charles Dickens' *Christmas Carol*, will have eaten bread contaminated with alum.

And the addition of alum not only deprived England's poorest population of necessary nutrients, but directly contributed to their ill-health. Alum's acidity could cause damage to the stomach, resulting, according to historian Dr Susannah Lipscomb, in bowel problems, constipation and chronic diarrhoea, which could be fatal in small children.[20] It was also held by some to be a contributory factor to a horrible disease that was rampant in urban Britain in the middle of the nineteenth century, one that twisted and contorted the human body, leaving its victims with 'a stunted figure, a large head, misshapen chest, tweeted long bones, and enlarged wrists and ankles'.[21] Incredibly and horrifically, up to sixty per cent of children in London suffered from rickets. Tiny Tim Cratchit likely had rickets along with TB.[22]

Among those who suspected that alum was a direct cause of rickets was Doctor John Snow, whose word carried weight because he had successfully reduced the number of cases of cholera in Soho after his campaign to close down a water pump which he had identified as a cause of the outbreak. He published his hypothesis about alum and rickets in England's most respected medical journal, the *Lancet*, in 1857, and he was indeed on the right track again. For while the direct cause of rickets is a lack of Vitamin D, the inflammation of the bowels caused by alum can result in the body absorbing less Vitamin D from digested food.

The government reacted in 1862 with the *Adulteration of Food Act* which proved ineffective, and a second act of parliament in 1872 which was equally so. It was not until the *Sale of Food and Drugs Act* of 1875 made the adulteration of food a crime in itself, without having to prove that the seller had knowledge of it, that the situation improved. One estimate is that the rate of adulteration in bread fell to 0.6% by 1888.[23]

But the effects of malnutrition had a long reach. When the British army came to recruit soldiers for the first Boer War in 1880, a massive 34 per cent or more of recruits were found to be unfit for military service, primarily and scandalously as a result of rickets or other poverty-related diseases.' Alum had played its part in this.[24]

⬩⬩❖⬩⬩

Our stories of alum are coming to an end, but before closing this chapter there is one omission I must put right.

Google tells me that I have used the word 'worker' fifty times in my manuscript so far – but unlike the entrepreneurs (such as the Zaccarias in Phocaea) or the mine owners (such as Judith Baker in Boulby), not one of my many alum workers in different ages and in different parts of the world has featured as an individual, with a life, a date of birth, a family or even a name of his own. Of course, alum workers throughout the world were largely illiterate and left no written record of life seen from their perspective. Yet, as Nobel Peace Prize winner Elie Wiesel says, we 'must not see *any* person as an abstraction'; every person deserves to be seen as 'a universe with its own secrets, with its own treasures, with its own sources of anguish and with some measure of triumph'. So, in honour of alum workers everywhere, let me try to put the flesh and bones on an alum worker who lived in Yorkshire in the nineteenth century. I can't claim to enter his secrets, but I can try to picture at least some aspects of the life he lived. [25]

Meet Thomas Temple, born around 1797 in the village of Skelton-in-Cleveland. Thomas was one of 512 children baptised between 1781 and 1801 in Skelton's All Saints Church, where the great and the good of the village had their own pews, with their names inscribed on the side panels. The Temples sat at the back, with the poor of the village, oblivious to the warnings written on large panels on the walls of the church ('The eyes of the Lord are in every place, beholding the evil and the good') because they could not read.

The usual pattern of employment in rural areas was that men were employed as agricultural labourers, while women were rarely in employment at all. But Skelton was an exception, because both men and women were employed in the alum mines to the west (Guisborough) and to the east (Loftus and Boulby). These workers often had a cow of their own, or a couple of sheep or pigs, as a buffer against difficult times. The animals grazed on areas of common land either side of the byways outside the village while their owners toiled in the alum works.

The early 1800s were, for many, difficult times. Food prices, already high as a result of the wars with France, rose still further following a spate of particularly cold winters. Then, in 1816, dust from the eruption of faraway Mount Tambora in Indonesia blacked out the summer sun completely, resulting in the coldest July on record and a very poor harvest. With food scarce, there was an increase in poaching, even though a law of 1800 punished poachers with whipping and hard labour. Two months of hard labour was the punishment for widow Knaggs, for milking a neighbour's cow and taking its milk.[26]

There were no schools for children like Thomas. The first school in Skelton opened in 1814, when he was 17. So the two topics (apart from the high cost of food and fodder) that likely dominated the talk he heard around him as he grew up maybe came as close as anything to constituting his formal education.

One topic was the fear of invasion by the French. The fear was widespread and real, more real than we might often imagine, knowing with hindsight that no invasion ever materialised. Local people were trained in a nationwide organisation of militias, as a line of defence if ever an invasion took place. Thus young Thomas will have seen men he knew marching in Skelton's local volunteer association, learning to shoot to kill. A day's walk away from Skelton is Hartlepool where, it is said, people hanged a monkey that had escaped from a sinking French ship, mistaking it for a Frenchman. The story is almost certainly a fabrication, but it does convey something of the hysteria that was rampant at the time.

The second topic, less dramatic but longer lasting, was the legal theft of the common land by the landed classes in a policy euphemistically called 'enclosure', which saw most of England's greens and wide grassy verges disappear into private ownership, behind walls and hedges. Enclosure hit different parts of England at different times, and it hit Cleveland from 1813. So in his teenage years, Thomas will have heard his parents and neighbours bitterly complain that they now had nowhere to graze their cow, or pig, or sheep. Most labourers had to sell or slaughter their animals, losing an all-important source of emergency income. They were powerless to do anything about it. This was the law of the land, and neither men nor women in Thomas's class had the right to vote.

Enclosure made it one notch harder for people like Thomas to avoid the workhouse in Skelton, or the bigger one that replaced it in Guisborough in 1838. Even worse was the dreaded House of Correction in Northallerton, the number of whose inmates grew alarmingly: 268 in 1824, 306 in 1833, 710 in 1843, 862 in 1853.[27] In 1821 the House of Correction installed a new instrument of brutal punishment: a treadmill on which gangs of prisoners worked six to ten hours a day, ten minutes on, five minutes off, climbing the equivalent of 5,000 feet (1,660 metres) every day. The House of Correction in Northallerton soon had five of them, and even that was apparently not sufficient: in 1845 it installed, extraordinarily, what was claimed to be the biggest (read: meanest) treadmill in the world. Ninety-four men worked it, treading side by side.[28]

How did Thomas Temple navigate through a life stacked against him without ending up in one of these institutions? The census returns show us how. He moved from place to place in search of work, taking on any job that was going.

In 1831 Thomas was living in Lythe, near Sandsend. By 1841 he was in Staintondale, working as an agricultural labourer. In 1851, still in Staintondale, he was an alum worker: with the nearest alum works being the Peak works in Ravenscar, Thomas will have had a daily three-mile walk to and from work. By 1861 he had moved again, to South

Waterfall, Guisborough. This was the year that Prince Albert died and Queen Victoria went into mourning, creating a boom in Yorkshire in the mining of a lustrous black gemstone called jet. And what had Thomas become at the age of 63? Sure enough, a miner in the jet industry. And ten years later, in 1871, 73-year-old Thomas was still working – as a jet labourer.[29]

It is a remarkable life of mobility, adaptability, guts and sheer hard physical labour. For it is of course perfectly possible, indeed likely, that Thomas lived at yet more addresses and had yet more different kinds of employment in the decades between the censuses. And yet along the way, Thomas managed to marry and bring up a family. He married Elizabeth (née Park) in 1827. She was nine years younger than he, only twenty-one when they married, and the young couple immediately experienced the tragedy and trauma of losing their first two children, Elizabeth in 1827 and her sister Jane in 1831.[30] Elizabeth then gave birth to six sons in a row between 1831 and 1845, four of whom became miners in the jet and ironstone industries that ultimately replaced the Yorkshire alum industry.[31] And Thomas and Elizabeth were granted the joy of grandparenthood.

In 1871 Thomas and Elizabeth were living in Belmangate, Guisborough, along with two of their sons, William aged 26, and Joseph, aged 36. Indeed all the Temple boys seem to have lived with their parents for many years. What was their home like? Was it like the living room – meaning, literally, the only room in which the family lived – that the Rev J C Atkinson saw in cottages in Skelton in the late 1830s: 'The floor was of clay and in holes, and around on two sides were the cubicles, or sleeping boxes … for the entire family. There was no loft above, much less any attempt at a 'chamber'; only odds and ends of old garments, bundles of fodder and things of that sort and in this den the occupants of the house were living.' We cannot know.

The older cottages that we see in Belmangate today look very quaint, but we should be wary of romantic fantasies about the past. A number of houses in Belmangate were 'cleared' by compulsory purchase order in 1960.[32] And the sanitary conditions in what is now one of the most picturesque villages in the area, Runswick Bay, were so atrocious that as late as 1866 the village was struck by cholera. 'We found the place in a most deplorable condition, beggaring description,' wrote the health inspectors who came to investigate the cause of the outbreak. 'A monster group of midden steads ... Many houses are going to decay without any attempt being made either to arrest or expedite the process ... The quantity of human excrement of children and adults, everywhere interrupting the passage of visitors, too truly attested the scantiness of privy accommodation ... Were fevers of every kind things to be desired, no better means could have been devised for securing a never-ending succession of them than is now practised by the natives.'[33]

And yet, despite conditions such as these, Thomas lived to a good old age. Which was, it turns out, less of an exception than I had anticipated. The death rate at childbirth and in early infancy was horrendous, and infectious diseases such as flu, diphtheria, measles, whooping cough and typhoid were all too often killers, yet a good many people, even among the poor, led long lives. Of the 799 people buried in Skelton during the first half of Thomas's life (1813-1852), 133 were aged 70-80, 109 were 80-90, and 19 were 90-100, and two, remarkably, were 101.

Thomas Temple, one-time alum worker, died aged 83 in 1880, and Elizabeth, mother of eight children, died the same year, aged 74. Both were buried in Guisborough, having shared and helped each other through the joys, the blows and the sorrows that came their way in fifty-three years of married life together.

Oh, and there is a postscript.

Through their second son Joseph, alum worker Thomas and his wife Elizabeth Temple were the great-great-great-great-grandparents of Kate who married William in 2011.

The effect of the alum industry on the Royal Family remains to be seen.

17

The End and Beyond

More than three thousand tons of alum were shipped from Whitby in 1841 – a colossal amount.[1] Twenty years later Walter White saw 'wagon-loads of alum' carried at the Boulby alum works while offshore 'more than a hundred vessels of different sizes' were in the business of servicing the local alum industry.[2]

To an outside eye then, the Yorkshire alum industry appeared to be going strong in the mid-nineteenth century. At the Great Exhibition in London in 1851, alum was displayed in the form of a tall crystalline pillar that towered over the people around it. Many stood and admired it, and some maybe, if they followed the advice of the photographer, educator and author Samuel Prout Newcombe for how children should investigate the exhibits at the Exhibition, licked it.[3]

But the Yorkshire alum industry was, in fact, in deep trouble. Significantly, the alum displayed at the Great Exhibition in London (and again at the Paris Universal Exhibition of 1878) was not from Yorkshire at all, but from new rival works.[4] At Hurlet and Campsie, just outside of Glasgow, Charles Macintosh, the man whose interest in chemistry and textiles would later lead him to invent the waterproof raincoat, had begun to manufacture alum by concentrating the alum liquor not in open pans, but in a boiler furnace – much as the Chinese had been doing for centuries at Wenzhou. This method enabled Macintosh to gain a greater yield of alum from an equivalent amount of shale and to undercut the manufacturers in Yorkshire. As a result, the price of alum fell to £12 per ton by 1835, and to £9 10s per ton by 1846, and soon Macintosh was supplying the dyers in and around Glasgow, including seventeen or so businesses in Paisley that specialised in a decorative tear-drop motif that has come to be called Paisley Pattern.[5]

The writing had been on the wall, in fact, ever since chemists began experimenting with an entirely new method of producing alum from sulphuric acid. France, with few reserves of alunite or alum shale to exploit, was at the forefront of this research and in 1787 the French chemist, Jean Antoine Chaptal, was the first to produce alum by exposing clay containing alumina (aluminium oxide) to the fumes produced in the manufacture of sulphuric acid. While Chaptal's method did not immediately catch on internationally, he was soon using it to satisfy the alum needs of the French textile industry, thereby reducing its traditional dependence on Tolfa.[6]

From 1846 the British chemist Peter Spence, building on Chaptal's ideas, produced alum even more cheaply and quickly at Pendleton in Lancashire by using sulphuric acid on cheap coal mining waste. This enabled him to churn out more alum in a month than Yorkshire's coastal alum industry could produce in a year, and by 1862 Spence was making half of all the alum produced in England.[7]

The final hammer blow came with the invention of synthetic dyes, which removed the very need for alum in the first place. The first such dye was marketed by William Perkin, a British chemist who discovered that a mixture of aniline extracted with alcohol not only produced an intense purple dye, but one that was resistant to washing and sunlight. It was, in other words, a dye that contained its own fixative – and from that moment alum's role as the mordant of choice in the world's textile industries was broken. More and more aniline dyes were perfected for different colours, and when they were brought into commercial use, alum had become largely redundant.

The alum industry had for centuries been a forerunner of the Industrial Revolution. When that revolution finally kicked in however, it initiated a pace of change with which the alum industry could not compete. Which explains why, when it came, the end of the alum industry was so swift, so thorough and so complete.

The alum works in Loftus still produced 411 tons of alum in 1866; by 1871 they had shut down completely. Guisborough and Sandsend had closed the year before. Judith Baker's former works in Boulby tried to limp on by branching out into Epsom Salts, a by-product of the alum industry. To no avail. Like Loftus, it closed in 1871. The tall chimney of its alum works was still standing in a photo taken in 1900; by 1910 all that remained were some odd bits of wall.[8]

It was the same story across Europe.

Prussia's largest alum works, the Ennert-Hardt works at Holzlar, in the Prussian-controlled Rhineland over the river from Bonn, were going strong in 1850, and a drawing of the complex in 1853 shows several large buildings and two tall chimneys. By 1875, it had closed.[9]

Further east, in the German province of Thuringia, the *Morassina* and *Jeremias Glück* alum mines[10] had made their owners, the Frege family of bankers, rich enough to cough up two-thirds of the 24 million francs ransom that Napoleon exacted from the city of Leipzig, and then, following Napoleon's defeat at Waterloo, to buy back Saxony's crown jewels in 1819.[11] But both mines had closed by 1860. Further east again, the alum mine in Plauen had thrown in the towel in 1826, despite its name *Ewiges Leben*, meaning 'eternal life'.[12]

In Sweden, the *Latorp* alum, opened in 1765, closed in 1869. The alum shale here had a strong orangy red colour however, so even in its death throes, the industry threw out yet another new use for alum: the burnt alum shale was used as a hard-top on tennis courts and athletics race tracks.[13]

Ubiquitously, inexorably, an industry that had long been so visible with the billowing smoke from its smouldering towers of shale, so penetrating with the obnoxious stench from its vats of urine, so greedy with its insatiable need for wood as fuel, so transformative of the landscape with its massive dumps of burnt shale, withdrew from people's sight and smell.

For centuries the alum industry had been desecrating the face of the earth. Now, suddenly, it had vanished from it.

In time too, it disappeared from people's consciousness, and it was as though the noise, the smell, the labour and the industry itself had never been. And the sites of the former alum mines have fared very differently.

When, for example, the Egyptian oasis city Kharga, which as we know supplied alum to Egyptian pharaohs, Roman and Byzantine Emperors and Muslim Caliphs, applied for UNESCO heritage status in 2015, its bid cited its temples, its Roman forts, its Christian Byzantine cemetery, its underwater aqueducts and its caravan routes, but made no mention of Kharga's alum mines at all.[14]

The spirit of the Genoese survives in the castles at Foça (Phocaea) and Feodosia (Caffa), even if much of what we see today is Ottoman. And Istanbul's grandiose Galata tower, built by the Genoese in 1348, remains a city icon to this day. But the alum mines of Phocaea and Koloneia, which to a great extent financed the Genoese empire by supplying alum across the eastern Mediterranean and as far away as Bruges and Southampton, survive only in the names. Şap Tepesi in Phocaea means 'alum hill', while, as we have seen, Sebinkarahisar translates as 'alum black fortress'. Only Kütahya is still in the alum business, with its Gediz mine producing some 20 000 tonnes of alum a year,[15] mainly for use in medicines and in the cosmetics industry. None worth mentioning is for the business of dyeing.

In Genoa itself, the streets of the old city centre are lined with magnificent palaces built by families like the Doria, the Spinola, the Lomellini and the Giustiniani, many of whom made their fortune in alum. Among the forty or so listed as UNESCO World Heritage sites, not one, but two were built by Tobias Pallavicino, the alum tycoon who defrauded the pope of his alum income at Tolfa. One is the ornate Palazzo Carrega-Cataldi in the via Garibaldi, the other is the palatial Villa Pallavicino delle Peschiere, famously described by Charles Dickens, who stayed in it on his tour of Italy in 1844, as 'something larger than Whitehall multiplied by four'.[16]

At Tolfa, the alum mines which were to have financed a crusade but funded the papacy instead, have bequeathed us the town of Allumiere, originally built for the alum workers, and more surprisingly perhaps, a rich mixed forest of beech, oak and chestnut that is unique so far south in Italy. It exists thanks to one of the earliest known examples of woodland management, introduced in 1608 to make sure that the Tolfa alum mines would not run out of burning material.[17]

A number of former alum mines across the world have found new life as tourist attractions. In Germany, the *Ewiges Leben* mine in Plauen maybe has eternal life after all, because you can visit it, along with the former *Jeremias Glück* and *Morassina* mines near Saalfeld. The *Morassina* mine is used for wedding ceremonies.[18] And over the border in Czechia, the flooding of an alum mine in Chomutov has created a lake (the *Kamencové jezero*) whose 1% alum content is so powerful that no plant or animal life can exist in it, earning it the name of Europe's 'Dead Sea'.[19]

Both Christina Piper's former alum mine at Andrarum[20] and the former Bulahdelah alum mine in New South Wales sport tourist trails that tell the story of their former industries, while the haunting ruins of Alunite, Utah, qualify it as one of the 'ghost towns' in America's Far West.[21] Easily the most complete alum complex to visit, however, is the Wenzhou alunite site in China, where the works and the accompanying alum workers' village of Fudewan morphed almost seamlessly from industrial production into industrial museum. With its serried ranks of chimneyed kilns, Wenzhou became the first site in China to be awarded UNESCO World Heritage status for its industrial heritage.[22]

Surprisingly, maybe, the remains of Britain's alum industry are much more difficult to find. The names of Alum Bay on the Isle of Wight and Alum Chine in Bournemouth are red herrings, recording the places where alum was desperately sought but never found. In Yorkshire, information panels along a National Trust trail at Ravenscar bravely try to explain the meaning of the odd bits of brick wall scattered bewilderingly in the fern and heather. Some tunnelling survives near Boulby, a path from Littlebeck leads to a cave that was once the entrance into an alum mine, and a car park in Sandsend, near Whitby, is – rather weirdly – accessed through the stone wall of what was once the alum factory. It does not add up to much.

But these are the wrong places to look. Lift your eyes, and the evidence of Yorkshire's former alum industry is monumental and unmistakable in the headlands that poke out into the sea. They should all be flat-topped, with a cliff face dropping at right-angles vertically down to the sea, but those at Skinningrove (Boulby), Kettleness, Sandsend, Saltwick Bay and

Ravenscar look as though some huge monster has taken giant bites out of them, leaving them with oddly lumpy silhouettes. The lumps and the dells between them are not grassy green, nor purple with heather, nor yellow with gorse, nor russet with faded fern, but milky grey from a thick coating of soft and wispy alum shale waste. This barren layer is so deep that vegetation struggles in vain to grow through it, and when you walk on its unrelenting grey surface, it gives way just a little. It feels, as a young Polish couple enthused, when I happened to meet them on the headland south of Skinningrove, as though you are walking on the surface of the moon.

But no monster took bites out of the cliffs, and this is of course not the moon. The strange outline of the cliffs and the thick carpeting of alum shale are the silent witnesses to the alum industry that existed in Yorkshire for two hundred and fifty years.

18

Alum, a Niggle and a Surprising Answer

Something has been niggling at the back of my mind while I have been writing this book, and the time has come to tackle it head-on.

All right, the stories of alum make sense in the context of the great global trade networks. Someone, many centuries ago, discovered that alum could be used to fix colour in cloth. This knowledge travelled the trade routes and was exploited in the world's great textile centres.

So far, so good. But people who never tapped into those wider trade networks will also have wanted colour in their cloth. So what did they do? Did they discover the secrets of alum for themselves? Or did they maybe find different mordants that could be used as effectively?

And how could I find out, given that these are the people who tended to leave little written evidence?

Fortunately, the German philosopher, explorer, scientist and man of letters Alexander von Humboldt came to my aid. He certainly knew all about alum. As the Inspector of Mines for the state of Prussia, he visited the *Morassina* alum mine in July 1792 with a view to increasing production. This was because King Frederick Wilhelm II's big plan to get people to grow more potatoes was backfiring disastrously owing to an outbreak of potato blight – and one of the chief weapons against potato blight was alum.[1]

So keen was Humboldt to help alum miners, in fact, that he almost died in an alum mine. He had invented a gadget designed to let miners' lamps shed light for longer if they found themselves trapped underground. In 1796 Humboldt tested his gadget in a far corner of the alum mine at Bad Berneck, near Bayreuth, where oxygen was scarce. The result? Humboldt's gadget kept a small flame alive, but Humboldt himself passed out for lack of oxygen. He only survived thanks to the prompt action taken by his guide.[2]

When, then, during his five-year journey through South America (1799-1804) Humboldt found indigenous people trading alum near Cumana on the northern coast of Venezuela, he knew exactly what the stuff should look like.

And it didn't.

It was neither alunite as in Tolfa, nor alum shale as in the Morassina mine. It was, as Humboldt recognised at once, a natural evaporate of alum like that which once formed around the oasis lake in Kharga.

And here is the point. The indigenous people of Venezuela were selling their alum to the dyers and shoemakers in the town of Cumana, who were using it to fix the colours in their cloth and leather. And as this trade was going on independently of the Spanish occupiers who were so ignorant of any local supply that they imported alum from Spain, this strongly suggests that the indigenous people had themselves hit on the mordant qualities of alum.[3]

Here, then, was the answer to my first question. The people seen by Humboldt were untouched by the world's great trade routes, so they had worked out for themselves, quite independently, that alum could be used as a mordant. And if they had done so, it is perfectly possible that other people had done so too. Indeed, in his book *Peruvian Natural Dye Plants*, Kay Antunez de Mayolo records dyers using alum in the Andes.[4]

Was it possible that other people had used a different substance as a mordant? The answer to this question had a surprise in store for me.

We know that there are materials other than alum – various leaves and barks, for instance – that will also work as mordants. Their effect is neither as efficient nor as long-lasting as alum, but they serve. People who used them will have had clothes whose colours were in no way as luminous as those worn in Beijing, Istanbul and Bruges, but this will not have kept them awake at night. The luminosity of their clothing was likely not high on their agenda.

Now, one of the plants that was widely used as a mordant was clubmoss (*diphasiastrum complanatum*). The Vikings, for example, used clubmoss to dye their cloth, and maybe even imported it into Britain, for archaeologists have found it at Viking dyeing sites in England and Scotland.[5] The Anglo-Saxons too, used clubmoss as a mordant before trade with the Mediterranean gave them the secret of alum. Investigations into fragments of a woman's woollen shawl woven in Finland at some time between 800 and 1300 AD have revealed that, there too, clubmoss was the mordant when the shawl was dyed. We know that clubmoss was used in this way from Poland to Iceland, and possibly by pre-European cultures in North America.

How good was clubmoss as a mordant? As it happens, an opportunity to find out came in 2015 after floods destroyed valuable textiles at the Jorvik Viking Museum in York. The Vikings, it was known, had originally used clubmoss as a mordant in Jorvik (the Viking name for York), so when the museum replaced the lost textiles, it bought coloured wool from a company that used clubmoss as the mordant. Just to be able to compare, however, the company treated one batch of yellow wool with alum instead, and the results were interesting. While alum undoubtedly created a yellow colour that was stronger and more vivid, the colour fixed by clubmoss was perfectly acceptable.[6]

And there is, it turns out, a good scientific reason for this. Clubmoss is what biologists call a bio-accumulator: a plant that, over time, builds up and stores a foreign substance within its organism. And the substance stored in the case of clubmoss is ... yes, you guessed it: alum.

Nor is clubmoss unique in this respect. Indonesian weavers used the bark of the symplocos tree as a mordant, and the symplocos tree, native to East Asia, is also a bio-accumulator of alum.[7]

Some people, then, who had no knowledge of the processing of alum and used mordants other than alum to fix colour in their cloth, had by trial and error come to use plants which, unknown to them, naturally contained alum. In other words, without knowing it they had been using alum all the time.

And that, for the alum enthusiast that I have become while writing this book, is just about as wonderful a conclusion as I could possibly have hoped for.

References

Chapter 1

1. Le Grand, H. E. (1984) 'Theory and Application: The Early Chemical Work of J. A. C. Chaptal', *The British Journal for the History of Science, 17*(1), p. 34. http://www.jstor.org/stable/4026015
2. De La Llave, R. C. (2015) 'Making and using alum in Hispanic craft recipe books from the fifteenth and sixteenth centuries', in *Icon* 21 pp.51-65. Accessed April 28, 2021. http://www.jstor.org/stable/24721693.
3. Glick, T. F. (2005) *Medieval Science, Technology, and Medicine: An Encyclopedia*, edited by Thomas F. Glick, Steven Livesey, Faith Wallis. New York: Routledge, entry for Alum https://books.google.co.uk/books?id=hr62AgAAQBAJ&pg=PT137&lpg=PT137&dq=alum+%22Byzantine+Empire%22&source=bl&ots=hgTHSDl82d&sig=ACfU3U3mYTPLUUsoqLvE-MOYdwQirJHm2w&hl=en&sa=X&ved=2ahUKEwjtk_Gv6NHpAhVHTxUIHY81AyoQ6AEwC3oECAkQAQ#v=onepage&q=alum%20%22Byzantine%20Empire%22&f=false
4. Gollhammer, J. *Handel und Bankwesen in Italien im Spätmittelalter (2.4 Handelswege)*, University of Salzburg, accessed 19/03/2021 at https://www.sbg.ac.at/ges/people/janotta/diehanse/page-se-arbeiten/searbeitgollhammer.htm
5. A richer red was produced from the dried bodies of a female louse in the Eastern Mediterranean, whose name *kermes vermilio* gives us our word vermilion.
6. Souza, G. R. (2009), *The French Connection: Indian Cottons and their early modern technology,* in *How India Clothed the World: The World of South Asian Textiles, 1500-1850*, edited by Giorgio Riello, & Tirthankar Roy (2009). Boston, Leiden: Brill, p. 352

7. Nørtoft, M. J. 'The red-blue conundrum an archaeo-linguistic approach to red dyes and blue flowers in prehistory', in *Archaeological Textiles Review* Nr 59, 2017, p. 45
8. Boisseuil, D. (2020) L'alun à la fin du Moyen Age: Nouvelles approaches, nouvelles perspectives in *Alum landscapes Archaeology of production and network economy*, edited by Luisa Dallai, Giovanna Bianchi, Francesca Romana Stasolla. All'Insegna del Giglio, Sesto Fiorentino, p. 21.
file:///C:/Users/User/Downloads/L_Dallai_G_Bianchi_F_Stasolla_a_cura_di.pdf
9. Roman Emperors from the reign of Alexander Severus (225–235 CE) onwards, including Byzantine Emperors.
Munro, John H. (2007) 'The Anti-Red Shift—To the Dark Side: Colour Changes in Flemish Luxury Woollens, 1300–1550', in *Medieval Clothing and Textiles Volume 3,* edited by Robin Netherton Gale and R. Owen-Crocker, The Boydell Press, Woodbridge, p. 57
https://www.economics.utoronto.ca/wwwfiles/archives/munro5/MedCloth04chapfourC.pdf
10. Setton, K. M (1976) *The Papacy and the Levant, 1204-1571: The fifteenth century*, p. 239
https://books.google.co.uk/books?id=0Sz2VYI0l1IC&pg=PA388&lpg=PA388&dq=Vatican+alum+mines&source=bl&ots=4zXk0x62uJ&sig =ACfU3U0QVyXs-hYw6VIXVQCBxpKtPz5BLA&hl=en&sa=X&ved=2ahUKEwjxqvXtss_pAhWmUxUIHX1KDQs4ChDoATAEegQIChAB#v=onepage&q=alum&f=false
11. The best wools were March wools (Welsh Marches) of Herefordshire and Shropshire; the Cotswolds wools of Gloucestershire, Worcestershire, and Oxfordshire.
Munro, (2007), p. 63
12. While the term alum now denotes a double sulfate of either aluminium and potassium or ammonium, or a mixture of both (Al2 (SO4)3 K2 SO4 24H2O; Al2 (SO4)3 (NH4)2 O4 24H2O), it was formerly also used for the so-called "natural alums," most of which, at the end of the Middle Ages, were found in sediments or concretions in the Mediterranean.

Dumolyn, J. and Lambert, B. (2018) A Chemical Compound in a Capitalist Commodity Chain: The Production, Distribution and Industrial Use of Alum in the Mediterranean and the Textile Centers of the Low Countries (Thirteenth-Sixteenth Centuries) in *Journal of Early Modern History* 22. 10.1163/15700658-12342574, p. 240

Chapter 2

1. Ohsiro, Michinori, (2008) 'Kharga Oasis and Thebes: The Missing Piece of the Puzzle in the Relocation of Amen Worship in the 27th Dynasty?', *Orient*, published by the German Orient-Institute in Berlin, Vol. XLIII, p. 80 https://www.jstage.jst.go.jp/article/orient/43/0/43_75/_pdf
2. Kharga and Dakhla are now two separate oases. In ancient times they were still linked and known as the Great Oasis.
3. Ohsiro, M. (2008), p. 80
4. Ohsiro, M. (2008), p. 77
5. Beadnell, H.J. L. *An Egyptian Oasis*, London, J. Murray, 1909, p. 92
6. Boozer, A. L. (2015) 'The Social Impact of Trade and Migration: The Western Desert in Pharaonic and Post-Pharaonic Egypt' Oxford Handbooks Online https://www.oxfordhandbooks.com/view/10.1093/oxfordhb/9780199935413.001.0001/oxfordhb-9780199935413-e-37
7. Roger S. Bagnall, R. S., Aravecchia, N., Cribiore, R,. Davoli, P., Kaper O. E., and McFadden, S. (2015) 'An Oasis city', chapter 6 published as part of the NYU Library's Ancient World Digital Library and in partnership with the Institute for the Study of the Ancient World (ISAW), p. 250
8. Ohsiro, M. (2008), p. 77
9. Natural alum came as crystals or in a wispy thread-like form. "In ancient times, alum ... was used as a generic term for impure mixtures of sulphates of iron and aluminium that occurred naturally".
   Bogensperger, I. (2015) 'Alum in Ancient Egypt: The written evidence', *Excavating, analysing, reconstructing Textiles of the 1st millennium AD from Egypt and neighbouring countries.* Proceedings of the 9th conference of the research group 'Textiles

from the Nile Valley' Antwerp, 27–29 Nov. 2015, edited by Antoine De Moor, Cäcilia Fluck and Petra Linscheid (Tielt, Belgium: Lannoo Publishers, 2017), p. 255

10. There were deposits of alunite at Dakhla and al-Shabb. Maciej Paprocki, (2019) *Roads in the Deserts of Roman Egypt: Analysis, Atlas, Commentary*, Oxford: Oxbow Books, p 48
11. Rickard, D. (2015) *Pyrite: A Natural History of Fool's Gold*, Oxford: OUP, p. 38
    https://books.google.co.uk/
12. *The Whatmans and Wove paper*, Appendix I, Part 1, http://www.wovepaper.co.uk/alumessay1.html
    Various words were used to describe alum and similar minerals, so we can't always know exactly which mineral is being referred to. The link between vitriol and alum is particularly unclear, e.g. a Sumerian compilation of minerals of 4000 years ago lists vitriols as mordants in the dyeing industry.
    Rickard, D. (2015) p. 36
13. Sapsford, M. (2008/9) , *The use of sodium salt deposits in medical and medically associated industries in Ancient Egypt.* PhD, Cranfield University, p. 42 & pp. 92-93
14. Giumlia-Mair, A. (2016) 'Technology transfer from ancient Egypt to the far east?', *The Silk Road, Volume 14*, p. 128 https://edspace.american.edu/silkroadjournal/wp-content/uploads/sites/984/2017/09/GiumliaMair_SR14_2016_126_146_Pl8.pdf
15. from about 1550 BCE
16. Smith, J. (2010) 'In search of key blue ingredient in ancient Egyptian pottery', *Science Daily,* Washington University in St. Louis, https://www.sciencedaily.com/releases/2010/03/100317121354.htm
    Jones, S. (2014) *Ceramics - Art or Science?* Section 9.34 New Kingdom Pottery https://www.ceramicsartorscience.co.uk/EicBookUserFiles/Ceramics%20-%20Art%20or%20Science%20-%20Dr.%20Stan%20Jones.pdf
17. Ikram, S (2011) *Discoveries in the Desert: The North Kharga Oasis Survey*, lecture accessed on YouTube https://www.youtube.com/watch?v=7UFvvjAxViQ

18. Michailidou, A. (1999) 'Systems of Weight and Social Relations of 'Private' Production in the Late Bronze Aegean', *From Minoan Farmers to Roman Traders: Sidelights on the Economy of Ancient Crete*, edited by Angelos Chaniotis.
Stuttgart: Franz Steiner Verlag, p. 99
https://books.google.co.uk/books?id=wnI_lvO0198C&pg=PA106&lpg= PA106&dq=Pylos+tablets+From+Minoan+Farmers+to+Roman+Traders:+Sidelights+on+the+Economy+of+Ancient+Crete+by+ Anna+Michailidou&source=bl&ots=kg8nh5a9Sk&sig=ACfU3U1Ee71NsXHqEVhXCmA7sn4rcD-x1w&hl=en&sa=X&ved=2ahUKEwj1oMLKqY_uAhXJz4UKHfwlC1IQ6AEwBHoECAEQAg#v=onepage&q=Pylos%20tablets%20&f=false
19. The mina was, like the pound, a unit of both weight and currency.
20. Sapsford, M. (2008/9), pp. 92-93
21. Nimrud is also known as Kalhu
22. Salvatore, G (2018) *Textiles in the Neo-Assyrian Empire: A Study of Terminology*, Berlin: De Gruyter p. 124
https://books.google.co.uk/books?id=4wJPDwAAQBAJ&pg=PA125&lpg=PA125&dq=alum+Egypt+desert&source=bl&ots=JOyp8TBVvx&sig=ACfU3U2YW07nXCoT42eUrSXVIDB0536N9A&hl=en&sa=X&ved=2ahUKEwiP9IzC34bqAhVNQRUIHQYNCjgQ6AEwDHoECAkQAQ#v=onepage&q=alum%20Egypt%20desert&f=false
23. Seaford, R. (2004) *Money and the Early Greek Mind.* Cambridge: CUP, p. 80
1000 talents = 26 tons:
Photos-Jones, E. (2018) 'From mine to apothecary: an archaeo-biomedical approach to the study of the Greco-Roman lithotherapeutics industry', *World Archaeology*, Volume 50:3, p 421
https://www.tandfonline.com/action/showCitFormats?doi=10.1080%2F00438243.2018.1515034
24. Bresson, A. (2019) *The making of the ancient Greek economy.* Princeton University Press, p. 355

25. Giumlia-Mair, A. (2005) 'Alum in Ancient Metallurgy', *L'alun de Méditerranée* edited by Philippe Borgard, Jean-Pierre Brun, Maurice Picon, pp. 335-341 https://books.openedition.org/pcjb/616?lang=en
26. According to Herodotus, it took them just seven days to march to Kharga from the Nile at Thebes. Ohsiro, M. (2008), p. 77
27. Wuttmann, M., Gonon, T., Thiers, C. (2000) 'The Qanats of 'Ayn Manawir (Kharga Oasis, Egypt).' First International Symposium on Qanat, Yazd, Iran May 8-11, 2000, Yazd, Iran. p.162. ffhalshs-02051616f
28. English, P. W. (1968) 'The Origin and Spread of Qanats in the Old World', *Proceedings of the American Philosophical Society, Vol. 112, No. 3, Published by: American Philosophical Society*, pp. 170-181 http://www.jstor.org/stable/986162
29. Beadnell, H.J. L. (1909) *An Egyptian Oasis*, London, J. Murray, pp. 222-223
30. 'Water Treatment in History: The Ancient World, as illustrated by Rome', *Association of Water Technologies.* https://www.awt.org/resources/seed-program/water-careers/water-treatment-in-history/
31. Oshiro, M, (2008), p. 77
32. Adams, C. (2013) 'Natural Resources in Roman Egypt: some aspects of extraction, transport and administration', pp. 6-7 https://livrepository.liverpool.ac.uk/3004639/1/Natural_Resources_BASP_FINAL.pdf
33. Uzel, I. (1999) 'The Ottoman-Turkish Dentistry', *XIII. Turkish History Congress Ankara* , p. 458 http://islamicmanuscripts.info/reference/articles/Cicek-2000-Civilisation-3/Cicek-2000-Civilisation-3-1-455-470-Uzel.pdf
34. Rome still had the considerable income from the alum mines on Lipari and Stromboli.
35. Schneider, J. (1978), *Peacocks and penguins: the political economy of European cloth and colors*, City University of New York Graduate Center in American Ethnologist, p. 427
36. Schneider, J. (1978), p 416

37. Guest, Peter. (2008). 'Roman Gold and Hun Kings: the use and hoarding of solidi in the late fourth and fifth centuries', *Roman Coins Outside the Empire: Ways and Phases, Contexts and Functions,* Moneta, Wetteren, Institute of Archaeology, University of Warsaw, p. 296
    accessed at file:///C:/Users/sydth/Downloads/RomanGoldHunKings%20(1).pdf
38. For questioning whether Christ could be both God and man.
39. Jacoby, D. (2005) 'Production et commerce de l'alun oriental en Méditerranée, XIe-XVe **siècles**', *L'alun de Méditerrannée,* edited by Philippe Borgard, Jean-Pierre Brun and Maurice Picon, Publications du Centre Jean Bérard, pp 219-267. https://books.openedition.org/pcjb/601
40. Alum was most frequently measured in Genoese 'cantara', similar to the English hundredweight, and equivalent to just under 50 kilos (47,6 kg). The Ottoman kantar was a bit heavier (57 kg).
    Aygün, C. Ö. (2020) 'The flesh-eating stone: alum mining and trade in Asia Minor', *Alum landscapes Archaeology of production and network economy*, edited by Luisa Dallai, Giovanna Bianchi, Francesca Romana Stasolla. All'Insegna del Giglio, Sesto Fiorentino, p. 177
    Accessed at ResearchGate file:///C:/Users/sydth/Downloads/IPaesaggidellallume%20(5).pdf
41. Galliker, J. L. (2014) 'Middle Byzantine Silk in Context: Integrating the Textual and Material Evidence', Centre for Byzantine, Ottoman and Modern Greek Studies Classics Ancient History and Archaeology College of Arts and Law University of Birmingham, p. 130. https://etheses.bham.ac.uk/id/eprint/5560/5/Galliker15PhD.pdf
42. Jacoby, D. (2005), pp. 219-267

Chapter 3

1. Wikipedia, History of paper, quoting: Buisseret, D. (1998) *Envisaging the City*. Chicago Press, p 12
2. I saw this method of making paper by hand demonstrated at the Paper Mill Museum in Amalfi in 2022.

3. Xuan paper is entered on the list of UNESCO's Intangible Cultural Heritage of Humanity.
4. In Europe, alum was first used in the sizing of paper in the 16th century.
   Brückle, I. (1993) 'The Role of Alum in Historical Papermaking', *Abbey Newsletter, Volume 17, Nr 4*, based on a talk presented at the IPC Conference 1992 in Manchester. https://cool.culturalheritage.org/byorg/abbey/an/an17/an17-4/an17-407.html
5. Cartwright, M. (2017) 'Paper in Ancient China', *Ancient History Encyclopedia*. https://www.ancient.eu/article/1120/paper-in-ancient-china/#:~:text=The%20invention%20of%20paper%20greatly,convenient%20to%20use%20and%20cheaper.&text=Paper%20was%20so%20highly%20valued,(618%2D907%20CE).
6. Cartwright, M. (2017) 'Paper in Ancient China'
7. McMullen, D. L. (2011) 'The Chinese Examination System in Dynastic China: Did it select the brightest and best?', *Sunway Academic Journal Volume 8*, p. 3 https://core.ac.uk/download/pdf/148366621.pdf
8. Han, J. (2015) 'Imperial Yellow: A Costume Colour at the Top of the Social Hierarchy', blog of the Asia Research Institute of University of Nottingham. https://theasiadialogue.com/2015/07/22/imperial-yellow-a-costume-colour-at-the-top-of-the-social-hierarchy/
9. Gu, C. (2020) 'History and Origin of First Patents (Monopolies) in Song Dynasty China 1113 AD: Salt licenses 盐引, Tea licenses 茶引, Alum licenses 矾引'
   https://www.linkedin.com/pulse/history-origin-first-patents-monopolies-song-dynasty-china-chen-gu
10. Walters, O. W. (1962) 'Early Indonesian Commerce and the Origins of Srivijaya', thesis submitted for the Degree of Doctor of Philosophy in the University of London (Ann Arbor: Proquest, 1962), p. 265
11. It is printed on seven panels of paper.
12. The book is displayed in the British Library, and may be viewed online at the British Library website (the Diamond Sutra, Final Section).

Boat paintings: Webber, P. (2004) 'A souvenir from Guangzhou', *Conservation Journal Issue 48, VAM* http://www.vam.ac.uk/content/journals/conservation-journal/issue-48/a-souvenir-from-guangzhou/

Chapter 4

1. Ogden, J. M. (1990) 'Gold jewellery in Ptolemaic, Roman and Byzantine Egypt', Durham theses, Durham University, p 55 http://etheses.dur.ac.uk/1457/2/1457_v2.pdf
   Oddy, A. (1988) 'Unlocking the secrets of gold antiquities', *Gold Bulletin* 21, pp. 29–36 https://doi.org/10.1007/BF03214665 page 31
2. Ogden, J. M. (1990), p 56
3. Rumo, M. and Martelli, M. (2014) 'Near Eastern origins of Graeco-Egyptian Alchemy', *Dahlem Seminar for the History of Ancient Sciences Vol. II., Chapter 4*, edited by Klaus Geus and Mark Geller, Max-Planck-Institut für Wissenschaftsgeschichte, p 42.
   Accessed via Researchgate, file:///C:/Users/sydth/Downloads/NearEasternOriginsofGraeco-EgyptianAlchemy-MPIGW2014%20(1).pdf
4. Ogden, J. M. (1990), p. 57
5. Sherwood Taylor, F. (1952) *The Alchemists*. London: The Scientific Book Club, p. 20. https://archive.org/stream/in.ernet.dli.2015.63938/2015.63938.The-Alchemists--Founders-Of-Modern-Chemistry_djvu.txt
6. Giumlia-Mair, A. (2005) 'Alum in Ancient Metallurgy', *L'alun de Méditerranée* edited by Philippe Borgard, Jean-Pierre Brun, Maurice Picon, p. 335-341
   https://books.openedition.org/pcjb/616?lang=en
7. Rumo, M. and Martelli, M. (2014), p. 49
8. Rumo, M. and Martelli, M. (2014), p. 42
9. Ogden, J. M. (1990), p. 57
10. The coins were exhibited at the Government Museum in Egmore, Chennai, in January 2011. As reported by the Times of India, 19 January 2011 http://timesofindia.indiatimes.com/

articleshow/7315542.cms?utm_source=contentofinterest&utm_medium=text&utm_campaign=cppst, accessed 12/01/2021

11. Giumlia-Mair, A. (2005), pp. 335-341
12. Al-Hassan, A. Y. 'Arabic Alchemy 'Ilm al-San'a: Science of the Art', *History of Science and Technology in Islam* http://www.history-science-technology.com/articles/articles%20 10.html
13. Wang, Q., Vlachou-Mogire, C., Gooch, M. and Cook, B. (2015) 'A metallographic study of some debased silver coinage of Henry VIII in The British Museum', *Technical Research Bulletin, Volume 9*, p. 61

    Accessed via ResearchGate at file:///C:/Users/User/Downloads/BMTRB9Wangetal.pdf
14. Ogden, J. M. (1990), p. 57
15. Needham, J. (1997) *Science and Civilisation in China.* Cambridge: CUP, p. 130
16. Giumlia-Mair, A. (2005), pp. 335-341
17. Butcher, K. (2020) *Introduction to Debasement: Manipulation of Coin Standards in Pre-Modern Monetary Systems.* Oxford: Oxbow Books, p. 11

    https://books.google.co.uk
18. Follett, K. (2020) *The Evening and the Morning.* Basingstoke: Macmillan. The alum mixture is here called 'vitriol', but as we have seen, the terms vitriol and alum are often used inter-changeably.
19. Wang, Q., Vlachou-Mogire, C., Gooch, M. and Cook, B. (2015), p. 61
20. McQ. Holmes, N.M. and Van Caelenberghe, H. (2014) 'Some Scottish 'Black Money' and its imitations found in the Netherlands and Belgium', *The British Numismatic Journal*, edited by Martin Allen, pp. 110 & p. 129. https://www.britnumsoc.org/publications/Digital%20BNJ/pdfs/2014_BNJ_84.pdf
21. Townrow, S. K. (2012) 'No one much cared: Specie Counterfeiting and Community Support in 1780s Hampshire County', *Historic Deerfield Summer Fellowship*, pp. 7, 10, 11, Appendix II and Appendix III.

http://stephanietownrow.com/wp-content/uploads/2020/02/Historic-Deerfield_Thesis_Krom-Townrow_2012.pdf accessed 27/01/21

Chapter 5

1. The original horses are now displayed inside the cathedral, to protect them from air pollution. The horses on the outside of St Mark's are replicas.
2. Walton, N. (2015) *Genoa, La Superba.* London: Hurst & Company, pp. 12-13
   Krueger, H. C. (1933) 'Genoese Trade with Northwest Africa in the Twelfth Century.' *Speculum, vol. 8, no. 3*, p. 377
   Accessed at *JSTOR*, www.jstor.org/stable/2848865. Accessed 26 Mar. 2021
3. Vincens, É. (1842) *Histoire de la République de Gênes*, Livre Premier, Chapter 2 http://www.gutenberg.org/files/18669/18669-8.txt
4. Menzel, U. (2007) 'Imperium oder Hegemonie? Folge 3: Genua und die mediterrane Weltwirtschaft 1261-1350', Technische Universität Braunschweig, p 34
5. Menzel, U. (2007), p. 35
6. Menzel, U. (2007), p. 35
7. Vincens, É. (1842), Livre Premier, Chapter 4
8. Vincens, É. (1842), Livre Quatrième, Chapter 1
9. David Jacoby notes that the ships that took on salt in Phocaea in the 20th century were about 600 to 800 tonnes: the tonnage had not significantly increased since the 14th Century.
   Jacoby, D. (2005) 'pp. 219-267
10. Guérin, S. M. (2017) 'Exchange of Sacrifices: West Africa in the Medieval World of Goods', *The Medieval Globe, Volume 3, Issue 2*, Arc Humanities Press p. 102 https://muse.jhu.edu/article/758477
11. A reminder that the Genoese cantara was equivalent to 47.65 kilos. Balard, M. (1978) *La Romanie Génoise*. Genoa: Atti della Societa Ligure di Storia Patria, p. 538

https://core.ac.uk/download/pdf/362206704.pdf
The alum was probably from Sijilmassa in Morocco.
Jacoby, D. (2005), pp. 219-267
Bijaya:
Baadj, A. S. (2013) 'The Struggle for North Africa between Almohads, Ayyubids, and Banū Ghāniya (Late Twelfth to Early Thirteenth Centuries A.D.)', thesis submitted at Department of Near and Middle Eastern Civilizations University of Toronto, p. 133

Sirte:
Braun, K, and Passon, J. (Editors) (2020) *Across the Sahara*, Springer, p. 225
file:///C:/Users/User/Downloads/2020_Book_AcrossTheSahara.pdf

Alexandria 1140:
Balard, M. (1978) *La Romanie Génoise*. Genoa: Atti della Societa Ligure di Storia Patria, pp. 719 and 770
Accessed at https://core.ac.uk/download/pdf/362206704.pdf

12. Zaimeche, S. (2017) 'An introduction to Early Islamic Society and Social Sciences', *Muslim Heritage*, Quoting Singer: *The Earliest Chemical Industry,* p. 85.
https://muslimheritage.com/an-introduction-to-early-islamic-society-and-social-sciences/#_ednref181
13. Briys, E. and Joos de ter Beerst, D (2006) 'The Zaccaria Deal: Contract and options to fund a Genoese shipment of alum to Bruges in 1298', presented at XIV International Economic History Congress, Helsinki, p. 51 https://cyberlibris.typepad.com/blog/files/iehcpaperbriysjoos.pdf
Fleet, K. H. (1993) 'Trade Relations Between the Turks and the Genoese, 1300-1453', School of Oriental and African Studies, University of London, p. 205 https://eprints.soas.ac.uk/34111/1/11015927.pdf
14. Menzel, U. (2007), p. 27

15. Lopez, R. S. (1964) 'Market Expansion: The Case of Genoa', *Journal of Economic History, Vol. 24, No. 4,* p. 449 http://www.cultus.hk/hist/readingsBaltic/Robert_Lopez_Market_Expansion_The_Case_of_Genoa.pdf
16. Jahnke, C., & Köhler, S. (2019) 'Vom Mittelmeer zum Hanseraum: Der Handel mit Rosinen, Korinthen und Zibeben', *Hansische Geschichtsblätter, 136(2018),* p. 39 https://curis.ku.dk/ws/files/250344021/04_koehler_jahnke_sonderdruck_hgbll18.pdf
17. Menzel, U. (2007), p. 19
18. The figure is given by the Catalan mercenary Ramon Muntaner, who raided Phocaea in 1307, but is reckoned to be too high by David Jacoby
    Briys, E. and Joos de ter Beerst, D (2006), p. 133
    Jacoby, D. (2005), pp. 219-267
19. Some claim that all Egyptian alum was natural alum. This claim is disputed by others.
    Jacoby, D. (2005), pp. 219-267
20. Alunite ($KAl_3(SO_4)_2(OH)_6$) had to be transformed into alum ($KAl(SO_4)_2.12\ H_2O$).
    Aygün, C. Ö. (2020), p. 179
21. Briys, E. and Joos de ter Beerst, D. (2006), p. 133
22. Picon, M. (2005) 'Des aluns naturels aux aluns artificiels et aux aluns de synthèse: matières premières, gisements et procédés', Publications du Centre Jean Bérard, OpenEdition Books , pp. 13-38 https://books.openedition.org/pcjb/553
    Levey, M. (1958) 'Alum in Ancient Mesopotamian Technology', *Isis, vol. 49, no. 2*, pp. 166–169. *JSTOR*, www.jstor.org/stable/226929. Accessed 2 Dec. 2020
23. Solomon, S.: Water (2010) 'The Epic Struggle for Wealth, Power, and Civilization', Part II, Chapter 7. Accessed 02/03/2021 at https://publicism.info/environment/water/8.html
24. Briys, E. and Joos de ter Beerst, D. (2006), p. 133
    Lopez, R. S. (1951) 'Majorcans and Genoese on the North Sea Route in the thirteenth Century', *Revue belge de Philologie et d'Histoire* , pp. 1169 and 1176

Accessed at https://www.persee.fr/doc/rbph_0035-0818_1951_num_29_4_2120

25. Balard, M. (1978), p. 547
26. Salonia, M. (2015) 'Genoese Economic Culture: from the Mediterranean into the Spanish Atlantic', University of Liverpool Thesis, p. 28 https://livrepository.liverpool.ac.uk/2009805/1/SaloniaMat_Jan2015_2009805.pdf
27. Muslim Al-Andalus in southern Spain would certainly also qualify as one of Europe's most advanced urban societies.
28. Lopez, R. S. (1951), 'Majorcans and Genoese on the North Sea Route in the thirteenth Century', p. 1176
29. Le Goff, J. (1989) *Kaufleute und Bankiers im Mittelalter.* Frankfurt/Main: Fischer Taschenbuch Verlag GmbH, p. 17 https://monoskop.org/images/8/82/LeGoff_Jacques_Kaufleute_und_Bankiers_im_Mittelalter_1993.pdf
30. Lopez, R. S. (1951), 'Majorcans and Genoese on the North Sea Route in the thirteenth Century', p. 1168
31. Heyd, W, (1861) 'Die italienischen Handelscolonien in Griechenland während der Regierungszeit der drei ersten Paläologen. (1261-1341)', *Zeitschrift für die gesamte Staatswissenschaft / Journal of Institutional and Theoretical Economics, Bd. 17, H. 3./4.* Pp. 459-460 Accessed at: https://www.jstor.org/stable/40736782
    Menzel, U. (2007), p. 41
32. Lopez, R. S. Market Expansion: The Case of Genoa, pp. 446-447 http://www.cultus.hk/hist/readingsBaltic/Robert_Lopez_Market_Expansion_The_Case_of_Genoa.pdf
33. Lopez, R. S. (1951), 'Majorcans and Genoese on the North Sea Route in the thirteenth Century', pp. 1167-1169

Chapter 6

1. The UNESCO list includes a few cathedral towers that were used for civic functions, e.g. as look out towers.
   https://whc.unesco.org/en/list/943/
2. The earliest evidence of alum imports to Bruges is in 1163.

Briys, E. and Joos de ter Beerst, D (2006), p. 50

3. Reynolds, R. L. (1930) 'Merchants of Arras and the Overland Trade with Genoa Twelfth Century', Revue belge de Philologie et d'Histoire, 9-2 pp. 495-533 https://www.persee.fr/doc/rbph_0035-0818_1930_num_9_2_6699
4. Prial, F. (1986) 'Portrait of a medieval city', New York Times, Sept. 7, 1986, Section 10, p. 15 https://www.nytimes.com/1986/09/07/travel/portrait-of-a-medieval-city.html
5. Stabel, P. (2011) 'Le goût pour l'Orient: Demande cosmopolite et objets de luxe à Bruges à la fin du Moyen Âge', *Histoire urbaine, 1 (nr 30)*, pp. 21-39 https://www.cairn.info/revue-histoire-urbaine-2011-1-page-21.htm
6. Liddy, C. D., Haemers, J. (2013) 'Popular Politics in the Late Medieval City: York and Bruges', *The English Historical Review, Volume 128, Issue 533*, pp. 771–805, https://doi.org/10.1093/ehr/cet107
7. Van Houtte, J. A. (1966) 'The Rise and Decline of the Market of Bruges', *Economic History Review New Series, Vol. 19, No. 1,* p. 37 https://onlinelibrary.wiley.com/doi/abs/10.1111/j.1468-0289.1966.tb00959.x
8. Kendall, C. E. (1964) 'Trade guilds that influenced the textile industry of Western Europe, 12th through 14th centuries', thesis submitted to the Faculty of the Graduate School, University of North Carolina at Greensboro, p. 96
9. Dumolyn, J. and Lambert, B. (2018), p. 243
10. Heers, M-L (1954) 'Les Génois et le commerce de l'alun à la fin du Moyen Age', *Revue D'histoire économique et Sociale* 32, no. 1, pp. 34-49. http://www.jstor.org/stable/24070811
11. Oldland, J. (2019) *The English Woollen Industry, c.1200-c.1560.* Abingdon: Routledge, accessed at https://books.google.co.uk
12. Munro, J. H. (1966) 'Bruges and the Abortive Staple in English Cloth: An Incident in the Shift of Commerce from Bruges to Antwerp in the Late Fifteenth Century', *Revue belge de philologie*

et d'histoire, tome 44, fasc. 4, pp. 1137-1159 www.persee.fr/doc/rbph_0035-0818_1966_num_44_4_2652

13. Munro, J. H. (1966), p. 1149
14. Wiesner-Hanks, M. (2012) 'Early Modern Europe 1450-1789', *Cambridge History of Europe*, p 223 accessed at: file:///C:/Users/User/Downloads/early-modern-europe-1450-1789_compress.pdf
15. The oarsmen were from the Dalmatian (Croatian) coast, and worked on Venetian ships. https://www.stnicholascenter.org/gazetteer/7097
16. England produced 35,000 sacks of wool, which is the produce of nine million sheep.
    Cucu, F. (2020) 'Markets and Regional Development in the Long-Run: Evidence from Medieval England and Wales', p. 10, note 18. Available at SSRN http://dx.doi.org/10.2139/ssrn.3939018
17. Roskell, J. S., Clark, L., Rawcliffe, C. (editors) (1993) Entry for Southampton in: The History of Parliament https://www.historyofparliamentonline.org/volume/1386-1421/constituencies/southampton
18. Mowat, R. B. (editor) (1937) *The Overseas Trade of Bristol.* Bristol Records Society's Publications Vol. VII, p 83 https://www.bristol.ac.uk/Depts/History/bristolrecordsociety/publications/brs07.pdf
19. Salonia, M. (2015), pp. 30-31
20. Roskell, J. S. , Clark, L., Rawcliffe, C. (editors) (1993) Entry for Southampton in: The History of Parliament https://www.hisoryofparliamentonline.org/volume/1386-1421/constituencies/southampton
21. Morton, A. D. (2014) 'The French Raid on Southampton 1338, Part One', *Journal of the Southampton Local History Forum No. 23*, pp. 27-30
22. Salonia, M. (2015), pp. 31-32
23. *Brande, N., Fennell, D., Price, O., Pugh, F., Ryan, K., Smith, M. and Tozer, I. (undergraduate History students at the University of Southampton) (2021),* 'Immigrants in Later Medieval Southampton', guest post at Historic Southampton: posted April 29 https://historicsouthampton.co.uk/medieval-immigrants/
24. https://www.stnicholascenter.org/gazetteer/7097

25. Southampton's Wool & Cloth Heritage, Posted on July 9, 2012, accessed at https://comestepbackintime.wordpress.com/2012/07/09/southamptons-wool-cloth-heritage/
26. *Brande, N., Fennell, D., Price, O., Pugh, F., Ryan, K., Smith, M. and Tozer, I. (undergraduate History students at the University of Southampton) (2021)*
27. Coleman, O. (1963) 'Trade and Prosperity in the Fifteenth Century: Some Aspects of the Trade of Southampton'. *Economic History Review, New Series, 16, no. 1* , pp. 18-21 Accessed May 12, 2021. doi:10.2307/2592513
28. Oldland, J. (2019)
29. Currency converter of the National Archives, accessed at https://www.nationalarchives.gov.uk/currency-converter/#currency-result
   Fryde, E. B. (1972) 'Anglo-Italian Commerce in the Fifteenth Century: Some Evidence about Profits and the Balance of Trade, *Revue belge de Philologie et d'Histoire*, pp. 349-353
30. Mackintosh, I. (2020) 'The Dyer's 1471 Charter', The Worshipful Company of Dyers.
   https://www.dyerscompany.co.uk/wp-content/uploads/2020/10/1471-The-Dyers-Charter.pdf
31. Hare, J. (2015) 'The cloth industry', in English Inland Trade 1430 - 1540 edited by Michael Hicks. Oxford: Oxbow books, p. 58
   https://books.google.co.uk

Chapter 7

1. The quality of alum rock in Koloneia was the highest in Asia Minor. Percentage of alum ore in the alum rock was 80% in Koloneia, 75% in Phocaea on the western coast of Anatolia and 60% in Kütahya (Kotiaeion, Cotyaeum).
   Aygün, C. Ö. (2020), p. 180
2. Now Sebinkarahisar in northeast Turkey. Koloneia is the Greek name, Colonna in Latin. The Greeks also used the name Mavrokatron, which means Black Castle. The Turks called it Karahisar (Black Castle in Turkish), and then Sapkarahisar or Sebinkarahisar to

distinguish it from other Karahisars. 'Sap' and 'sebin' are Turkish for 'alum'.
Bryer, A. A. M. (1982) 'The Question of Byzantine Mines in the Pontos: Chalybian Iron, Chaldian Silver, Koloneian Alum and the Mummy of Cheriana,' *Anatolian Studies*, *vol. 32,* p. 148 *JSTOR*, www.jstor.org/stable/3642677. Accessed 1 Dec. 2020.

3. Miynat, A. (2017) 'Cultural and socio-economic relations between the Turkmen states and the Byzantine Empire and West with a corpus of the Turkmen coins in the Barber Institute coins collection', Thesis submitted to the University of Birmingham, p. 79. https://etheses.bham.ac.uk/id/eprint/7411/4/Miynat17PhD.pdf
4. Aygün, C. Ö. (2020), p. 177
5. Fleet, K. H. (1993), p. 213
   Aygün, C. Ö. (2020), p.180
6. Balard, M. (1978), p. 776
7. Jacoby, D. (2005), pp. 219-267
8. Aygün, C. Ö. (2020), p. 176
9. Barker, H. (2014) 'Egyptian and Italian Merchants in the Black Sea Slave Trade, 1260-1500', submitted in partial fulfillment of the requirements for the degree of Doctor of Philosophy in the Graduate School of Arts and Sciences Columbia University, p. 164
10. Frankopan, P. (2015) *The Silk Roads.* London: Bloomsbury, p. 177
11. Slater, E. (2006) 'Caffa: Early Western Expansion in the Late Medieval World, 1261-1475', *Review (Fernand Braudel Center), 29*(3), p 275. Retrieved April 21, 2021, from http://www.jstor.org/stable/40241665 p. 275
    Pegolotti quote:
    Frankopan, P., (2005) *The Silk Roads.* London: Bloomsbury, p. 184
12. Other bulk goods in the Black Sea region were grain and salt.
13. Jacoby, D. (2005), p. 147
14. Bryer, A. A. M. (1982), pp. 133-150
15. Khvalkov, I. A. (2015) 'The Colonies of Genoa in the Black Sea Region: Evolution and Transformation, Volume 1', Thesis submitted for the degree of Doctor of History and Civilization of the European University Institute Florence, p. 425 https://cadmus.eui.

eu/bitstream/handle/1814/40744/2015_AlexandrovitchKhvalkov_Vol_I.pdf?sequence=1&isAllowed=y

16. Slater, E. (2006), p. 272
17. Wheelis, M. (2002) 'Biological Warfare at the 1346 Siege of Caffa'. https://www.ncbi.nlm.nih.gov/pmc/articles/PMC2732530/ + https://discover-ukraine.info/places/crimea/feodosiya/554
18. Balard, M. (1978), p. 545
19. Khvalkov, I. A. (2015), p 13, based on: Reinhard, A Short History of Colonialism (Manchester/New York, Manchester University Press, 2011), p. 8.
    https://cadmus.eui.eu/bitstream/handle/1814/40744/2015_AlexandrovitchKhvalkov_Vol_I.pdf?isAllowed=y&sequence=1
    Thomas Kirk points out that Caffa was not a colony if we define colonies as being run by a state. Caffa was run by a private company and later by a bank, the Casa di San Giorgio.
    Kirk, T. (2020) 'The Republic of Genoa and its maritime Empire' in Thomas Kirk: Empires of the Sea. Leyden Brill 2020, p. 154. Accessed at: https://brill.com/view/book/edcoll/9789004407671/BP000007
20. Slater, E. (2006), p. 274
21. Bocharov, S. G. (2017) 'Ancient Economic and Social Concepts in the Genoese Gasaria Region', *European Research Studies Journal Volume XX, Special Issue*, pp. 199-207), https://core.ac.uk/download/pdf/162326627.pdf
22. Balard, M. (1978), p. 586
23. Balletto, L. (2017) 'L'émigration féminine de Gênes au Proche-Orient au xve siècle', *Migrations et diasporas méditerranéennes (X-XV siècles,* edited by Michel Balard and Alain Ducellier, pp. 297-311
    https://books.openedition.org/psorbonne/2242?lang=en
    De Roover, R. (1963) 'The rise and decline of the Medici Bank 1397-1494', *Harvard Studies in Business History XXI,* Harvard University Press, Cambridge, Massachusetts London, England, p. 58
    https://www.gwern.net/docs/history/1963-deroover-theriseanddeclineofthemedicibank.pdf

24. Fleet, K. H. (1993), p. 133
25. Slater, E. (2006), p. 272
26. Barker, H. (2014), p. 144
27. Setton, K. M. (1976) *The Papacy and the Levant, 1204-1571: The thirteenth and fourteenth centuries*, American Philosophical Society, p 272
https://books.google.co.uk
28. Heers, M-L (1954), p. 36
29. Barker, H. (2014), pp. 12-18
30. Özkutlu, S. (2014) 'Medieval Famagusta: socio-economic and sociocultural dynamics (13th to 15th Centuries)', thesis submitted to The University of Birmingham, p. 67. Accessed 12/04/2021 at https://etheses.bham.ac.uk/id/eprint/6111/1/Ozkutlu15PhD.pdf
31. The 1359-1363 sales were in Tana. Slaves for sale in Genoa and Venice were mainly aged 15-25.
Barker, H. (2014), pp. 130, 170, 176
32. Khvalkov, I. A. (2015), p. 450
33. Dash, M. (2015) Slavery on the Steppes: Finnish children in the slave markets of medieval Crimea
https://mikedashhistory.com/2015/01/15/blonde-cargoes-finnish-children-in-the-slave-markets-of-medieval-crimea/
34. Khvalkov, I. A. (2015), page. 437
35. Barker, H. (2014), p. 3
36. Verlinden, C. (1965) Review of *Gênes au XVe siècle* by Jacques Heers, *Revue belge de Philologie et d'Histoire*, p. 157, accessed at: https://www.persee.fr/doc/rbph_0035-0818_1965_num_43_1_2566_t1_0155_0000_2
37. Briys, E. and Joos de ter Beerst, D. (2006), p. 40
Barker, H. (2014), p. 81
38. Stanziani, A. (2018) 'Bondage: Labor and rights in Eurasia from the sixteenth to the early twentieth centuries', *International Studies in Social History, No. 24, Berghahn Books, New York*, p. 76
https://doi.org/10.2307/j.ctt9qcm9z
Rohan, P. (2015) The Genoese Levantine colonies at the birth of Ottoman imperial power: a framework for inquiry, thesis submitted

to the Graduate School of Social Sciences of Istanbul, Sehir University, p. 18

39. Barker, H. (2014), p. 113
40. Slater, E. (2006), p. 276
41. *Stello*, A. (2017) *La traite d'esclaves en Mer Noire (première moitié du xve siècle)*, Collection de la Casa de Velázquez, OpenEdition Books , pp. 171-179
42. Rohan, P. (2015), p. 21
43. Barker, H. (2014), pp. 90-91
44. Jacoby, D. (2005), paragraph 100
45. "Male slaves were also sent to the alum mines of Genoa at Focea." Stanziani, A. (2018), p. 76

    Zimmer, E. M. (2006-2018) Lederberg Memorial Website, Continuity and Medieval Colonization, quoting: Verlinden, C., 'The Beginnings of Modern Colonization: Eleven Essays with an Introduction', Cornell Univ. Press, 1970, p. 31. http://www.estherlederberg.com/Eugenics%20(Anecdotes)/Genoa.html
46. Barker, H. (2014) 'Egyptian and Italian Merchants in the Black Sea Slave Trade, 1260-1500', submitted in partial fulfilment of the requirements for the degree of Doctor of Philosophy in the Graduate School of Arts and Sciences Columbia University, p. 166
47. Stanziani, A. (2018) 'Bondage: Labor and rights in Eurasia from the sixteenth to the early twentieth centuries', *International Studies in Social History, No. 24*, Berghahn Books, New York, NY, https://doi.org/10.2307/j.ctt9qcm9z

    p. 76

    Slavery and Bondage in Central Asia and Russia from the Fourteenth to the Nineteenth Century, p. 76, accessed at: https://www.berghahnbooks.com/downloads/OpenAccess/StanzianiBondage/StanzianiBondage_03.pdf
48. Wheelis, M. (2002), Historical Background to the siege of Caffa
49. Frankopan, P. (2015) *The Silk Roads*. London: Bloomsbury, p. 188
50. Balard, M. (1978), p. 868
51. Balard, M. (1978), p. 531

52. In 1352. *Trading Posts and Fortifications on Genoese Trade Routes from the Mediterranean to the Black Sea* (2020) http://whc.unesco.org/en/tentativelists/6468
53. https://discover-ukraine.info/places/crimea/feodosiya/554
54. Bocharov, S. G. (2017), pp. 199-207
55. Balard, M. (1978), p. 854
56. At the end of the 14th century, Genoese Giovanni Demorede appears to have had the monopoly over the alum production.
    Dağlı, M. (2012) 'Kütahya in the 18th Century: Transformation or the Persistence of the Old Order?', dissertation submitted at the University of California, Berkeley, p. 46
    Aygün, C. Ö. (2020), p. 175
57. Khvalkov, I. A. (2015), p. 425
    Jacoby, D. (2005), pp. 219-267, paragraph 76
58. Jacoby, D. (2005), pp. 219-267, paragraphs 55, 56 & 64
    Arvanitidou. I. P. (2020) 'Alum mines in medieval Greece', *Alum landscapes Archaeology of production and network economy*, edited by Luisa Dallai, Giovanna Bianchi, Francesca Romana Stasolla. All'Insegna del Giglio, Sesto Fiorentino, pp. 197 and 198
59. Balard, M. (1978), p. 775
    Tiddeman, M. M. S. (2016) 'Money Talks: Anglo-Norman, Italian and English language contact in medieval merchant documents, c1200-c1450, Volume 1', thesis presented at Aberystwyth University, p 49
60. Verlinden, C. (1965), p. 156
    Heers, M-L (1954), p. 44
61. Rakova, S. (2008) 'Genoese in Constantinople', *Encyclopaedia of the Hellenic World, Constantinople,* section 5: Genoese-Ottoman Relations http://constantinople.ehw.gr/Forms/fLemmaBody.aspx?lemmaid=11933

Chapter 8

1. Battle of Manzikert, 1071, near Lake Van in the far east of Anatolia.
2. The Ottomans are named after Osman I. The 's' in Osman was pronounced a bit like a voiced 'th'. Taking the 'th' as a 't' and

inserting a second syllable to separate the two consonants, the Italians made the name into *Ottomani*, and the English (and French) followed suit. Germans, in contrast, talk about the *Osmanen*.

3. Jacoby, D. (2005), pp. 219-267, paragraph 59
4. Sapes was subjugated to the Ottomans in 1361. Like the Sebum in Sebum-Karahissar, the name Sapes is derived from the Turkish word for alum. Baira Clio: General information on the Municipality of Iasmos, in Regional Guide of E. Macedonia and Thrace, 10/03/2007 http://www.xanthi.ilsp.gr/cultureportalweb/print.php?article_id=468&lang=en&print_mode=article
   Gattilusio:
   Heers, M-L. (1954), p. 32
5. The Kütahya area included Gediz.
   Dağlı, M. (2012), p. 45
6. Jacoby, D. (2005), pp. 219-267, paragraphs 60, 80 and 85
7. Better known in Britain as Tamburlaine.
8. Jacoby, D. (2005), pp. 219-267, paragraph 85
9. Dağlı, M. (2012), pp. 26 and 47
10. Many families took on the surname Giustiniani to indicate their membership of the mahona.
11. There were silver and gold ducats, and ducats struck in Chios and Phocaea were of less value than those struck in Venice, so the value of a ducat is hard to pin down. But as Marie-Louise Heers points out, the figure of 2000 ducats shows us how the Ottomans applied pressure on the Genoese: the sum was trebled to 6000 ducats in 1453, raised to 10 000 ducats in 1456, and increased to 30 000 ducats in 1457.
    Heers, M-L. (1954), p. 50
12. Jacoby, D. (2005), pp. 219-267, paragraph 86
    Mercan, F. Ö. (2016) 'The Genoese of Pera in the Fifteenth Century' in: *Living in the Ottoman Realm.* Indiana
    University Press, p. 45 https://books.google.co.uk
13. Jacoby, D. (2005), pp. 219-267, paragraph 86
    Fleet, K. H. (1993), p. 224

14. Siege of Constantinople 1422, on the Byzantine Battles website, accessed 08/05/2021 at https://byzantium.gr/battle.php?byzbat=c15_02
15. Khvalkov, E. (2018) *The colonies of Genoa in the Black Sea region.* New York: Routledge, pp. 398-403
16. Khvalkov, I. A. (2015), p. 476
17. Heers, M-L. (1954), p. 37
18. For ducats, see note 223 above.
19. Wright, C. (2014) *The Gattilusio Lordships and the Aegean World 1355-1462* . Leiden: Brill, p. 182, accessed 02/06/20121 at https://books.google.co.uk
20. Khvalkov, I. A. (2015), p. 472
    Turetken, M. (2019) 'How it happened: Ottomans conquer Istanbul', Anadolu News Agency, Eti Mahallesi GMK Bulvarı No:132 Çankaya Ankara, Accessed at: https://www.aa.com.tr/en/turkey/how-it-happened-ottomans-conquer-istanbul/1492258
21. Mercan, F. Ö. (2016) , p. 48
22. Heers, M-L. (1954), p. 36
23. Phocaea was farmed out to two Venetians, Bartolomeo Zorzi and Girolamo Michiel, in 1463. They were farmed out again in 1475 but we do not know who to.
    Jacoby, D. (2005), pp. 219-267, paragraphs 109 and 111
24. Heers, M-L. (1954), p. 44
25. Heers, M-L. (1954), p. 50
26. Heers, M-L. (1954), pp. 51-53
    Venice: Jacoby, D. (2005), pp. 219-267, paragraph 105
27. Jacoby, D. (2005), pp. 219-267, paragraph 104
    Setton, K. M. (1976), pp. 239-240
28. Khvalkov, I. A. (2015), p. 472
29. Balletto, L. (2017), pp. 297-311
30. Carr-Riegel, L. (2021) 'Italian Traders in Poland 1300-1500'. Thesis submitted to Central European University, Budapest, p. 249 file:///C:/Users/sydth/Downloads/carr-riegel_leslie%20(2).pdf
31. Venice's connection to Tana survives in the name of one of Venice's canals, the Fondamenta de la Tana.

32. Bryer, A. A. M. (1982) p. 149
33. Vryonis, S. Jr. (1965) 'Byzantium: its internal history and relations with the Muslim World', *XII. Seljuk Gulams and Ottoman Devshirmes, in: Der Islam, XLI, Berlin* http://macedonia.kroraina.com/en/svp/svp_12.htm
34. Bryer, A. A. M. (1982) p. 148
35. Khvalkov, I. A. (2015), pp. 475-476
36. Barker, H. (2014), p. 190
37. The very last Genoese possessions to fall to the Ottomans were Monkastro and Kiliya in Moldavia in 1484.
    Bocharov, S. G. (2017), p. 205
38. Jacoby, D. (2005), pp. 219-267, paragraph 69
39. Benes,, C. E. (2018) *A companion to Medieval Genoa*. Leiden: Brill, p 416, accessed at https://books.google.co.uk
40. Jacoby, D. (2005), pp. 219-267, paragraph 108

Chapter 9

1. Schneider, J. (1978), p. 419
2. Schneider, J. (1978), p. 419
   Joanot Valero's Dyeing Manual of 1497, as quoted in
   De La Llave, R. C. (2015), p. 57
3. When boiled with a dye, the sulphate atoms in the alum react with the water to form sulphuric acid. In the 21st century, 52 of 61 workers still using traditional methods to dye cloth with alum in India developed respiratory problems within ten years. Alum is now classified as a moderately hazardous material, with no more than half a milligram of alum dust allowed per cubic metre of air in places where alum is used.
   De Rosen, E. (2018) 'The economic fate of urban settlements in Rhomanian Boeotia, Thessaly, and Western Macedonia (783-1204)', thesis submitted to the University of Birmingham, pp. 81-82
4. Glick, T. F. (2005), entry for Alum
5. Barlee, R. (2001) 'Aluminium Tannages', *Skin Deep Volume 11*, J. Hewit & Sons Ltd, 12 Nettlehill Road, Houstoun Industrial Estate, Livingston, West Lothian, EH54 5DL, Scotland

https://www.hewit.com/skin_deep/?volume=11&article=2

6. Schneider, J. (1978), p. 419
7. Schneider, J. (1978), p. 420
8. Schneider, J. (1978), p. 427
9. Schneider, J. (1978), p. 425
   Munro, J. H. (2007) 'The Anti-Red Shift—To the Dark Side: Colour Changes in Flemish Luxury Woollens, 1300–1550', in *Medieval Clothing and Textiles Volume 3,* edited by Robin Netherton Gale and R. Owen-Crocker, The Boydell Press, Woodbridge, p. 92
   https://www.economics.utoronto.ca/wwwfiles/archives/munro5/MedCloth04chapfourC.pdf
10. Galliker, J. L. (2014), p. 131
11. De Roover, R. (1963), p. 152
12. Leix. A. (1937) 'Dyeing and Dyers' Guilds in Mediaeval Craftsmanship', *Ciba Review 1: Medieval Dyeing, Basel* http://www.elizabethancostume.net/cibas/ciba1.html
13. Leix. A. (1937)
14. Jacoby, D. (2005), pp. 219-267, paragraph 110
15. Heers, M-L (1954), p. 52
16. Heers, M-L (1954), pp. 51-53
17. Fleming, P. (2013) *Time, Space and Power in Later Medieval Bristol*. Bristol: University of the West of England, p. 98 accessed at https://uwe-repository.worktribe.com/output/924983/time-space-and-power-in-later-medieval-bristol
    Jenks, S., (2006) *Robert Sturmy's commercial expedition to the Mediterranean (1457/8)*. The Bristol Record Society's Publications, Volume 58, p. 3
    http://www.bristol.ac.uk/Depts/History/bristolrecordsociety/publications/brs58.pdf

Chapter 10

1. Guenster, A. and Martin. S. (2015) 'A Holy Alliance: Collusion in the Renaissance Europe Alum Market', *Review of Industrial Organization 47(1),* section 3.1. https://www.researchgate.net/

publication/277881715_A_Holy_Alliance_Collusion_in_the_Renaissance_Europe_Alum_Market

2. Chaunu, P. (1964) Review of Jean Delumeau, 'L'Alun de Rome, XVe-XVIe siècles', Annales 19-4 p. 763
   https://www.persee.fr/doc/ahess_0395-2649_1964_num_19_4_421211
3. Guenster, A. and Martin. S. (2015), section 3.4
4. Stasolla, F. R. (2020) 'The allumiere of the Tolfa district: archaeology and economical network', and
   Vallelonga, F. (2020) 'The settlement of La Bianca; the first village of miners?', *Alum landscapes Archaeology of production and network economy*, edited by Luisa Dallai, Giovanna Bianchi, Francesca Romana Stasolla. All'Insegna del Giglio, Sesto Fiorentino, pp. 52 and 68.
   Jones, A. (2016) 'How a Pope Saved a Forest', *Philanthropy, United Nations*. Accessed at https://impakter.com/how-a-pope-saved-a-forest/
5. Ekelund, R. B, Tollison, R. D., Anderson, G. M., Hébert, R. F., Davidson, A. B. (1996) *Sacred Trust: The Medieval Church as an Economic Firm*, Oxford: OUP, p. 37 https://books.google.co.uk
6. Jacoby, D. (2005), pp. 219-267, paragraph 104
7. Quoted in: Setton, K. M. (1976), p. 239
8. Guenster, A. and Martin. S. (2015), section 3.4
   Quotation from:
   Penn, T. (2012) *Winter King: Henry VII and the Dawn of Tudor England*. New York: Simon and Schuster, p. 251
9. Setton, K. M. (1976), pp. 239-240
10. Jacoby, D. (2005), pp. 219-267, paragraph 104
    Bartolini, C. (2012), posted on Maestronet Forums 6 May 2012
    https://maestronet.com/forum/index.php?/topic/326061-history-of-alum/
11. Ekelund, R. B, Tollison, R. D., Anderson, G. M., Hébert, R. F., Davidson, A. B. (1996), p. 41, note 15.
12. Setton, K. M. (1976), p. 23
13. In 1476 Pope Sixtus IV, who disliked the Medici, took the Tolfa contract from them and awarded it instead to their rivals, the Pazzi,

following which the Tolfa contract was awarded to a Genoese consortium. So the Genoese were back in control of alum after all. Guenster, A. and Martin. S. (2015), p. 10

14. Wainwright, M. C., and Michelson, E, (2021) *A Companion to Religious Minorities in Early Modern Rome.* Leiden: Brill, pp. 276 and 310
file:///C:/Users/User/Downloads/ebin.pub_a-companion-to-religious-minorities-in-early-modern-rome-9789004443495.pdf
15. Justice, G. (2011) 'The Role of Indulgences in the Building of New Saint Peter's Basilica', (2011). Master of Liberal Studies Theses, 7, p. 30 http://scholarship.rollins.edu/mls/7
16. This was despite the fact that canon law deemed money made from monopolies to be *turpe lucrum* – dirty money.
De Roover, R. (1963), p. 155
17. Boisseuil, D. (2020) p.21
18. Jacoby, D. (2005), pp. 219-267, paragraphs 110 and 111
19. De Roover, R. (1963), p. 156
20. Guenster, A. and Martin. S. (2015), section 4.1
21. De Roover, R. (1963), p. 156
Guenster, A. and Martin. S. (2015), section 3.4
22. De Roover, R. (1963), p. 157
23. Dumolyn, J. and Lambert, B. (2018), p. 249
Walsh, R. J., (2005) *Charles the Bold and Italy (1467-1477): Politics and Personnel.* Liverpool University Press, p. 87
https://books.google.co.uk
24. Jacoby, D. (2005), pp. 219-267, paragraphs 112 and 113
25. Özkutlu, S. (2014), p. 162
26. De Roover, R. (1963), p. 161
27. Venice: 1486-1490, in Calendar of State Papers Relating To English Affairs in the Archives of Venice, Volume 1, 1202-1509. Feb. 19. MSS. in St. Mark's Library, No. 177 to Pope Innocent VIII, pp. 159-203
Originally published by Her Majesty's Stationery Office, London, 1864. https://www.british-history.ac.uk/cal-state-papers/venice/vol1/pp159-203

28. Penn, T. (2012), p. 203
29. Rohan, P. (2015), pp. 72-73
30. Soly, H, (1974), p. 804
31. Penn, T. (2012), pp. 251-252
32. Soly, H, (1974), p. 805
33. Penn, T. (2012), p. 252
34. Penn, T. (2012), pp. 202-203
    Setton, K. M. (1976), pp. 44 and 49
    Diviš, L. (2016) 'Henry the VII: A Renaissance Ruler?', Bachelor's Thesis, Thomas Bata University of Zlin, pp. 33 and 38-40
35. Penn, T. (2012), p. 204
    Zapf. K. (2020) 'Some Papal Bull: 16th Century Alum trade and English Royal Autonomy', Sound Ideas, University of Puget Sound: Summer Research 379, p. 20. https://soundideas.pugetsound.edu/summer_research/379
36. Bottin, J. (2014) 'Commerce, finances et pouvoir: la redistribution des aluns méditerranéens dans l'Europe du Nord-Ouest et en France au XVIe siècle', *Open Edition Journals: Mélanges de l' école française de Rome (Moyen Age)*, section 8 https://doi.org/10.4000/mefrm.1647
37. Zapf. K. (2020) p. 24
38. 50,878 standard cloths in the first six years of Henry VII's reign, to 81,835 in the last six – an increase of 61%. Standard cloths were in theory 24 yards in length, two in breadth. In practice they varied enormously, so the figures are rough.
    Ramsey, P. (1953) 'Overseas Trade in the Reign of Henry VII: The Evidence of Customs Accounts'. *Economic History Review, New Series, 6, no. 2,* p. 177.
    https://ehs.org.uk/article/overseas-trade-in-the-reign-of-henry-vii-the-evidence-of-customs-accounts/
39. Diviš, L. (2016), p. 40
40. Ross. J. (2012) Henry VII, Pretenders and Rebellions, historytoday.com
    https://history-groby.weebly.com/uploads/2/9/5/6/29562653/henry_vii_pretenders_and_rebellions.pdf,

41. Ekelund et al. estimates the income from the alum concession at 100,000 florins in 1473. They quote Ludwig Pastor's data in History of the Popes V. III, as reporting the entire papal income of the year as being 300,000 florins.
    Stephenson, H. D. Jr. (2015) 'Unlucky in affairs of business.... Turning Points in the life of Lorenzo de Medici', submitted for the degree of Master of Arts in the Graduate Liberal Studies Program in the Graduate School, Duke University, p. 9
    https://dukespace.lib.duke.edu/dspace/bitstream/handle/10161/11158/Unlucky%20in%20affairs%20of%20business%20(final%20version).pdf;sequence=1
    Ekelund, R. B, Tollison, R. D., Anderson, G. M., Hébert, R. F., Davidson, A. B. (1996), p. 41
42. Chaunu, P. (1964), pp. 764-765
    One consequence was that about 30 000 cubic metres of wood were needed as fuel for manufacturing the alum.
    Jones, A. (2016)
43. Justice, G. (2011), p. 77
    Patenden, M., (2019) 'The Roman Curia', in *A companion to early modern Rome, 1492-1692*, Leiden: Brill, 2019) p. 54, https://books.google.co.uk
44. Stinger, C. L. (1998) C. L. *The Renaissance in Rome.* Indianapolis: Indiana University Press, p. 20
45. Setton, K. M. (1976), p. 49
46. Guenster, A. and Martin. S. (2015), section 5.2
    Rowland, I. D., (1984) 'Some Panegyrics to Agostino Chigi', *Journal of the Warburg and Courtauld Institutes*, vol. 47, 1984, p. 194. *JSTOR*, www.jstor.org/stable/751447. Accessed 20 Aug. 2021
47. The villa is the Villa Farnesina in Rome, and the Chigi Chapel is in the church of Santa Maria del Popolo.
    Durant. W., 'The Renaissance: The Story of Civilization, Volume V', Simon and Schuster ebook. Chapter XVIII, section VIII
    Masinter, M. (1993) 'Agostino Chigi's Villa Suburbana, a Roman lifestyle (1510-1520), submitted in partial fulfillment of the requirements for the degree of Master of Arts in the History of

Decorative Arts, The Cooper-Hewitt Museum and the Parsons School of Design file:///C:/Users/User/Downloads/MasinterAgostino_ChigiVillaNA7595.R64%20M39%201993%20CHMThesis.pdf

48. Guenster, A. and Martin. S. (2015), section 3.4
49. Ekelund, R. B, Tollison, R. D., Anderson, G. M., Hébert, R. F., Davidson, A. B. (1996), p. 37 and note 14 on p. 41
50. Cronin, V. (1969) *The Flowering of the Renaissance*. London: HarperCollinsPublishers, Chapter 3
51. Justice, G. (2011), p. 77
52. Justice, G. (2011), p. 32
53. Charles L. Mee, Jr. *White Robe, Black Robe,* New York: G.P. Putnam's Sons, p 176.
    Quoted in Ginny Justice: Justice, G. (2011), p. 77
54. Cronin, V. (1969), chapter 3
    Justice, G. (2011), p. 53
55. Encyclopedia Britannica, Volume 12, p. 269. Entry: Indulgences. Accessed at National Library of Scotland website: https://digital.nls.uk/encyclopaedia-britannica/archive/192941299?mode=transcription
56. Justice, G. (2011), pp. 27 and 55
57. Postrel, V. (2019) 'Not even a pope can maintain a monopoly', *Reason Magazine*, April Issue
    https://reason.com/2019/03/23/not-even-the-pope-can-maintain/

Chapter 11

1. To Castile in 1478, and later to Aragon and Portugal.
2. All Spanish Muslims were expelled in 1609.
3. Mindel, N. 'Don Joseph Nasi - Duke of Naxos', *Hebrew Word of the Week, Kehot Publication Society* https://www.chabad.org/library/article_cdo/aid/111925/jewish/Don-Joseph-Nasi-Duke-of-Naxos.htm
4. Birnbaum, M. D. (2013) *The long journey of Gracia Mendes*, Open books, Central European University Press, chapter 7, paragraph 63. https://books.openedition.org/ceup/2139?lang=en

5. Constance H. Rose. C. R., (1970) 'New Information on the Life of Joseph Nasi Duke of Naxos: The Venetian Phase', *The Jewish Quarterly Review, Vol. 60, No. 4*, pp. 333-334 https://www.jstor.org/stable/1453557
   Halil Inalcik, H., 'Turkey in Europe: A Historical Perspective', Bilkent University, Ankara, section 2 http://sam.gov.tr/pdf/perceptions/Volume-II/march-may-1997/HalilInalcik1.pdf
6. German wiki, Joseph Nasi, quoting Calimani, R., *Storia del ghetto di Venezia*, p 108
   Birnbaum, M. D. (2013), chapter 7, paragraph 122.
7. Mindel, N. 'Don Joseph Nasi - Duke of Naxos'
8. Jews paid property and poll taxes, for instance.
   Birnbaum, M. D. (2013), chapter 7, paragraph 10
9. Birnbaum, M. D. (2013) chapter 7, note 2.
10. 40,000:
    Iorga, N. (1971) 'Byzance après Byzance', *Association internationale d'études du sud-est européen, comité national roumain, Bucarest,* p. 51 https://upload.wikimedia.org/wikipedia/commons/7/73/Nicolae_Iorga_-_Byzance_apr%C3%A8s_Byzance_-_Continuation_de_l%E2%80%99Histoire_de_la_vie_byzantine.pdf
    50,000 and 250,000: Birnbaum, M. D. (2013) chapter 7, paragraph 44.
11. Birnbaum, M. D. (2013) chapter 7, paragraph 45.
12. Birnbaum, M. D. (2013) chapter 7, paragraph 50.
13. Faroqhi, S. (1979) 'Alum production and alum trade in the Ottoman Empire (about 1560-1830), *Wiener Zeitschrift für die Kunde des Morgenlandes, Vol. 71*, p. 156
    http://www.jstor.org/stable/23858910
    Sünnetçioglu, H. E. (2013) 'Hear the Other Side: The meaning of the Battle of Lepanto (1571) among late 16th century Ottoman historians', thesis at the Central European University, Budapest, p. 24 file:///C:/Users/User/Downloads/sunnetcioglu_halil.pdf
14. Simon, B. (1984) 'Contribution à l'étude du commerce vénitien dans l'Empire Ottoman au milieu du XVIe siècle (1558-1560)', *Mélanges de l'école française de Rome*, p. 990.

https://www.persee.fr/doc/mefr_0223-5110_1984_num_96_2_2778?q=Joseph+Nasi

15. Birnbaum, M. D. (2013) chapter 7, paragraph 60.
16. 400,000: Birnbaum, M. D. (2013) , chapter 7, paragraph 58.
    80: Iorga, N. (1971), p. 51
17. The Venetians paid the Sultan 1.320.000 akçe for the right to farm for three years. For comparison, the revenue of the governor of the province of Anatolia was one million akçe, while the building of the Suleiman Mosque in Istanbul in the 1550s is reputed to have cost 59 million akçe.
    Dağlı, M. (2012), p.47
    https://en.wikipedia.org/wiki/Ak%C3%A7e
18. Arbel, B. (1995) Trading Nations: Jews and Venetians in the Early Modern Eastern Mediterranean. Leiden: Brill, p. 115
    https://books.google.co.uk
19. Faroqhi, S. (1979), p. 162
    Arbel, B. (1995), p. 107
20. This was despite Venice sharing in the naval victory over the Ottomans at Lepanto in 1571. Venice was too weak to follow up on this victory and the Ottomans lived to win another day.
21. Faroqhi, S. (1979), p. 161
    Dağlı, M. (2012), p. 118
22. Dağlı, M. (2012), p. 47
23. Sebinkarahissar: Evliya Efendi (1850) *Narrative of travels in Europe, Asia and Africa in the 17th century*, Volume II. London: William Allen, p. 208 https://www.gutenberg.org/files/54255/54255-h/ 54255-h.htm
    Gediz: Aygün, C. Ö. (2020), p. 175
    dressing the horse's wound: Çelebi, E., (2011) *An Ottoman Traveller: Selections from the Book of Travels*. London: Eland Publishing Limited, p. 222 file:///C:/Users/User/Downloads/evliya-%C3%A7elebi-selections.pdf
24. Yaman, B. (2016) 'Producing soot ink by traditional methods and results of its application in the Ottoman book arts',
    International Journal of Business, Economics and Law, Vol. 9, Issue 5 (Apr.), pp. 91, 95 and 96
    https://www.ijbel.com/wp-content/uploads/2016/06/KLiISC_39.pdf

Chapter 12

1. Tu, H-Y. (2012) 'The Pursuit of God's Glory: Francis Walsingham's Espionage in Elizabethan Politics, 1568-1588', PhD thesis at the University of York History, pp. 88-89
2. Chaunu, P. (1964), p. 764
3. The Church of England considers itself a bridge between the Catholic and Protestant churches. Here I am simplifying by dividing western Europe in two: the Catholic south and Protestant north.
4. Schneider, J. (1978), p. 419
   Zapf. K. (2020) pp. 22 and 27-30
5. Potter, D. *Henry VIII and Francis I: The Final Conflict, 1540-47*. Leiden: Brill, p. 249
   https://books.google.co.uk
6. Rhine and Tyrol: Picon, M. (2005) pp. 13-38, section 2.2.4
   Schwemsal: Turton, R. B. (1938), *The Alum Farm*. Whitby: Horne & Son, p. 4
7. Plauen: https://www.plauen.de/Tourismus-Kultur-und-Freizeit/Service-und-Reiseinformationen/ Wegweiser-Barrierefrei-/index.php?object=tx,3317.1&ModID=9&FID=3317.2500.1&NavID= 3317.238&La=1
   Saalfeld: https://www.bergmannsverein-erfurt.de/main.php?unmenue=bergbau_frueher_thueringen&content=saalfeld
8. Beckmann, J. (1784) *Beyträge zur Geschichte der Erfindungen, Volume 2*. Leipzig: Verlag Paul Gotthelf Kummer, p. 141
9. Bauer (whose name means farmer in German) often latinised his name to 'Agricola'.
   Cotterill, M. (2016) 'Early British Alum – People, Quarries, Industry', section 16
   http://iwnhas.org/wp-content/uploads/2015/02/Early_British_Alum_People_Quarries_Industry_Pages1-84.pdf
10. Görkau: https://www.sudeten-bayreuth.de/staedte/goerkau.html; Chomutov: https://bookpub.club/book-read/sagenbuch-des-erzgebirges-1595539456593x981207731099866400
11. Gendorf's first alum works were in Kuttenberg (Kutna Hora in Czech), and as from 1544 200 km further west in Tschachwitz (Čachovice), near Kaaden (Kadan).

Pick, F. (1914) 'Zur Geschichte der Böhmischen Industrie in der Neuzeit;, *Vierteljahrschrift für Sozial- Und Wirtschaftsgeschichte, vol. 12, no. 4*, pp. 593–611, http://www.jstor.org/stable/20724874

12. Turton, R. B. (1938), *The Alum Farm*. Whitby: Horne & Son, pp. 7-8.
    Loss of £6000:
    Bagwell, R., (1885) *Ireland under the Tudors with a Succinct Account of the Earlier History, Vol 1.* London: Green and Co https://books.google.co.uk
13. Appleton, P. (2018) *A forgotten industry*. Skelton: Boroughgate Books, p. 15
14. Soly, H. (1974), p. 802
    Zapf. K. (2020), p. 13
    Spain: In 1562 Spain produced 26 000 cantara (12 000 at Mazarron, 8000 in nearby Lorca and 6000 at Cartagena.
    De la Llave, R. C., Silva A. F. and Espinach, G. N. (2015) *L'alun de la Péninsule ibérique durant la période mediévale (Royaumes de Castille et d'Aragon)*, Open Books, Publications du Centre Jean Bérard, pp. 125-137
15. Tobias Pallavicino's palace is now known as the Carrega-Cataldi Palace, nr 4 Via Garibaldi.
    Hollington, M. and Orestano, F. (2009) *Dickens and Italy*, Newcastle upon Tyne: Cambridge Scholars Publishing, Introduction p. xviii
16. Munro, J. H. (1966) 'Bruges and the Abortive Staple in English Cloth: An Incident in the Shift of Commerce from Bruges to Antwerp in the Late Fifteenth Century', p. 1149
17. Bonte, P., Porqueres i Gené, E., Wilgaux, J. (2017) *L'argument de la filiation*, Open Books, Editions de la Maison des sciences de l'homme, p. 434
    https://books.openedition.org/editionsmsh/8236?lang=en
18. http://www.treccani.it/enciclopedia/orazio-pallavicino_(Dizionario-Biografico)/
    Bottin, J. (2014), section 28
19. De Divitis, G. P. (1997) *English Merchants in Seventeenth Century Italy*, Cambridge: CUP, pp. 5-6

https://books.google.co.uk
In 1577 four merchantmen owned by Hawkins and his brother were chartered to ship alum from Civitavecchia.
Morgan, B. (2007) 'Hawkins, Sir John (1532–1595)', *Oxford Dictionary of National Biography*
https://www.oxforddnb.com/display/10.1093/ref:odnb/9780198614128.001.0001/odnb-9780198614128-e-12672

20. Bottin, J. (2014), table 1.
21. Chauviré, R. (1957) Review of *An Elizabethan : Sir Horatio Palavicino* by Laurence Stone, *Revue Historique, vol. 218, no. 1*, p. 141. *JSTOR*, www.jstor.org/stable/40948894. Accessed 23 Aug. 2021
Villani, S. (2014) 'Pallavicino, Horace', Biographical Dictionary of Italians, Volume 80
https://www.treccani.it/enciclopedia/orazio-pallavicino_(Dizionario-Biografico)/
22. Soly, H. (1974), p. 837
23. Soly, H. (1974), p. 837
24. Chauviré, R. (1957), pp. 141-143
25. Bonte, P., Porqueres i Gené, E., Wilgaux, J. (2017), p. 434
Ungerer, G. (1974) *A Spaniard in Elizabethan England: The Correspondence of Antonio Perez's exile, Volume 1,* London: Tamesis Books, p. 454
https://books.google.co.uk
26. MacGregor, N. (2014) *Shakespeare's Restless World.* London: Penguin, p. 96
Mantel, H. (1992) 'Blame it on the Belgians', London Review of Books, Vol. 14, No 12
https://www.lrb.co.uk/the-paper/v14/n12/hilary-mantel/blame-it-on-the-belgians
27. Villani, S. (2014)
28. Horatio received denizenship, a form of British nationality, in 1585.
29. Soly, H. (1974), p. 838
30. Soly, H. (1974), p. 838
31. Munro, J. H. (2011) 'Usury, Calvinism, and Credit in Protestant England: from the Sixteenth Century to the Industrial

Revolution,' *Working Papers tecipa-439, University of Toronto, Department of Economics*, Abstract.

32. Soly, H. (1974), p. 837
33. Soly, H. (1974), p. 839
34. De la Llave, R. C., Silva A. F. and Espinach, G. N. (2015) p. 125-137
35. Turton, R. B. (1938), p. 5
36. Bonte, P., Porqueres i Gené, E., Wilgaux, J. (2017), p 434
37. Soly, H. (1974), p. 839
38. Cotterill, M. (2016), section 13
39. Stone, L. (1948) 'The Anatomy of the Elizabethan Aristocracy', *The Economic History Review, vol. 18, no. 1/2*, pp. 5, 6, 15 & 17 JSTOR, www.jstor.org/stable/2590261. Accessed 25 Aug. 2021
40. Quoted in:
    John Cooper, J. (2012) *The Queen's Agent*. London: Faber & Faber, p. 324

Chapter 13

1. Bigland, J. (1812) *The Beauties of England and Wales, Or, Original Delineations, Vol. 16,* London: Longman and Co, pp. 322-323
   https://babel.hathitrust.org/cgi/pt?id=njp.32101073814996&view=1up&seq=353
2. Cotterill, M. (2016), section 36: Whitby Port and the alum industry
3. Turton, R. B. (1938), pp. 37-39
   Cotterill, M. (2016), section 4
4. Turton, R. B. (1938), pp. 64-68
5. A Lord Eure may have had an interest in alum springs said to exist at Oakengates in Shropshire
   'Swallow Head Farm: Wider Historical Context', with reference to document NYCRO MIC 3289/2066 in North *Yorkshire County Record Office*. https://www.maltonbuildingsgroup.com/history/old_malton/history_old_malton_swallow_head_management_plan_2.pdf

Later workings:
https://www.teeswildlife.org/what-we-do/past-projects/alum-alchemy-and-ammonites/places-to-visit/stoupe-brow-alum-works/

6. Cotterill, M. (2016), section 42 A: Allum-Garth: an early Yorkshire alum works?
   Young, G. (1817) *A History of Whitby*. Edinburgh: Longman, pp. 760-761 https://books.google.co.uk/
7. Turton, R. B. (1938), p. 1
8. Turton, R. B. (1938), pp. 71, 111 and 114
9. 1800 tons was estimated home demand.
   Appleton, P. (2018) p. 31
   But Turton records (1606) annual import of alum was 4200 tons.
   Turton, R. B. (1938), p. 73
10. Appleton, P. (2018) p. 26
    W. Hyde Price, W. H. (1906) *The English Patents of Monopoly*. Boston and New York: Houghton, Mifflin & Company, pp. 82-101 https://archive.org/details/englishpatentsm01pricgoog/page/n12/mode/2up?view=theater
11. Appleton, P. (2018) pp. 27 and 31
12. Soly, H. (1974), p. 840
13. 1st year losses: Turton, R. B. (1938), p. 76
    2nd year losses: Appleton, P. (2018) p. 27
14. For the process of making alum from alum shale:
    White, W. (1861) *A month in Yorkshire*. London: Chapman and Hall, pp. 97-103 https://www.gutenberg.org/files/35933/35933-h/35933-h.htm
    Turton, R. B. (1938), pp. 102-107
    Appleton, P. (2018) pp. 55-69
15. White, W. (1861), pp. 97-103
16. Some of these stones were taken by ship to Newcastle and sold as building material.
    Quinn, K. (1977) 'Some early examples of plant development and process control in the chemical industry', thesis presented for the degree of Master of Science, University of Durham, p. 44 http://etheses.dur.ac.uk/7448/1/7448_4513.PDF

17. Turton, R. B. (1938), p. 107
    + https://www.english-heritage.org.uk/visit/places/whitby-abbey/history-and-stories/history/
18. Turton, R. B. (1938), pp. 111-113
19. Turton, R. B. (1938), p. 78
20. Historic England Research Records: Sandsend Alum Quarries. https://www.heritagegateway.org.uk/Gateway/Results_Single.aspx?uid=1118514&resourceID=19191
21. Quinn, K. (1977), p. 23
22. https://web.archive.org/web/20110716030245/http://www.fortunecity.com/greenfield/ecolodge/25/alum.htm
23. The *Allome Ayme* goes by different names (Allome An, Allome Anna) in different accounts.
    Chaloner: Turton, R. B. (1938), p. 76
    Allomes Ayme: Turton, R. B. (1938), p. 108
    Take the piss:
    Hart-Davis A., Troscianko E. (2006) *Taking the Piss: a potted history of pee.* Stroud: The History Press, pp. 72-74
    Preference for labourers' urine + mixing with water:
    Quinn, K. (1977), p. 23
24. Rickard, D. (2015) p. 41
    For a detailed description of the chemical reactions involved, see Peter Appleton's excellent book, *A forgotten industry, pages 55 to 69. (Boroughgate Books, 2018)*
25. Barber, T. (2019) 'The History of Alum Production and its links to the Yorkshire Coast', Yorkshire's Best Guides: Blog, June 7 https://yorkshiresbestguides.co.uk/the-history-of-alum-production-and-the-links-to-the-yorkshire-coast/
    The figures vary wildly: one ton of alum required (1) up to 100 tons of shale (Whatmans), or (2) 20 tonnes of shale (paper of the British Geological Survey), or (3) at least 50 tons of shale (Susan Beaumont).
    1 Whatmans and Wove Paper. http://www.wovepaper.co.uk/alumessay2.html
    2 Hobbs, P. R. N., Entwisle, D. C., Northmore, K. J., Sumbler, M. G., Jones, L. D., Kemp, S., Self, S., Barron, M., and Meakin, J. L.

(2012) 'Engineering Geology of British rocks and soils', Lias Group. *British Geological Survey, Nottingham, UK.* (OR/12/032). http://earthwise.bgs.ac.uk/index.php/OR/12/032_Industrial_applications

3 Beaumont, S. L. (2019) 'Female Enterprise and Entrepreneurship in North East England (1778-1801)', submitted in partial fulfilment of the requirements for the degree of Doctor of Philosophy at Newcastle University, p 155
https://theses.ncl.ac.uk/jspui/bitstream/10443/4515/1/Beaumont%20S%20L%202019.pdf

26. Historic England website https://historicengland.org.uk/listing/the-list/list-entry/1019093?section=official-listing
27. Turton, R. B. (1938), p. 85
28. Turton, R. B. (1938), pp. 79 and 85
29. This was certainly the case in the 19th century. We do not know for sure when the practice began.
    Appleton, P. (2018) p. 82
30. Robert Turton: The Alum Farm (Whitby: Horne & Son, 1938), pages 99, 101
31. Turton, R. B. (1938), p. 127
    Swallow Head Farm. Wider Historical Context, page 26
32. Roberts, R. J. (2012) 'Two mean fellows grand projectors': The self-projection of Sir Arthur Ingram and Lionel Cranfield, Earl of Essex, 1600-1645', thesis submitted in partial fulfilment of the requirements of Teesside University for the degree of Doctor of Philosophy, pp. 277, and 288-289
    https://research.tees.ac.uk/ws/portalfiles/portal/8058571/254593.pdf
33. Turton, R. B. (1938), p. 2
34. Swallow Head Farm. Wider Historical Context, page 26
35. Appleton, P. (2018) p. 84
36. Turton, R. B. (1938), pp. 107, 113
37. Johnson, B. (2009) 'Chapter 6: Whitby West Pier to Blackhall Rocks (Block 1 NMP)', North East Rapid Coast Zone Assessment, Archaeological Research Services Ltd, p. 96
    https://www.archaeologicalresearchservices.com/wp-content/uploads/2018/01/Chapter-6-Whitby-West-Pier-to-Blackhall-Rocks.pdf

38. White, W. (1861), p. 106
39. Appleton, P. (2018) pp. 55-56
40. Whitby Magazine, October 1827, p. 321 https://www.google.co.uk/books/edition/The_Whitby_Magazine_and_Monthly_Literary/1EC55S9gumkC?hl=en&gbpv=1&dq=Ship+%22Endeavour,+Laverick%22&pg=PA320&printsec= frontcover
41. Vol. III, No. 5. May 1, 1845
http://www.stokesleypride.org.uk › history group
42. Turton, R. B. (1938), pp. 156-157
43. Chemical Trade Journal Dec 10, 1904, p. 544. https://www.google.co.uk/books/edition/The_Chemical_Trade_Journal_and_Oil_Paint/yi1KAQAAMAAJ?hl=en&gbpv=1&dq=Runcorn+Australian+alum&pg=PA544&printsec=frontcover
44. The original patentees, whose contracts were to have lasted 31 years, were offered compensation.
45. Turton, R. B. (1938), p. 98
46. Turton, R. B. (1938), pp. 118-119
47. Zekonyte, K. (2018) 'Projectors in Seventeenth Century England and their relevance to the field of Project Management', thesis submitted at the University of Brighton for the degree of Doctor of Philosophy March 2018, p. 127
https://cris.brighton.ac.uk/ws/portalfiles/portal/4785598/PhD_thesis_Kristina_Zekonyte_final.pdf
48. Turton, R. B. (1938), p. 168
49. Alum quarrying, The effect of extractive industry on the Yorkshire Coastal Landscape
https://web.archive.org/web/20110716030245/http://www.fortunecity.com/greenfield/ecolodge/25/alum.htm
50. Favier, J. (1965) Review of *Une ferme pendant trois siècles: l'alun de Rome* by Delumeau, J. , *Journal des Savants* pp. 590-597
https://www.persee.fr/doc/jds_0021-8103_1965_num_3_1_1116
Liège: Harsin, P. 'Etudes sur l'histoire économique de la principauté de Liège particulièrement au XVIIe siècle: Chapitre 1' in *Industries Liégeoises*
http://www.chokier.com/FILES/INDUSTRIE/Histoire_Economique_Liege_XVII-Harsin.html

51. Barker, R. (2011) *The Rise of an Early Modern Shipping Industry: Whitby's Golden Fleet, 1600-1750*. Martlesham: Boydell Press, p. 83 https://www.google.co.uk

Chapter 14

1. Feng, S., Qian, W., Sanchiz, J. M. C., White, R & Robinson, M. (2020) 'Continuity and Change in the Alum Industry: A Technological Approach to Wenzhou Alum Mine (Zhejiang Province, China, 14th-21st Century)', *Industrial Archaeology Review, 42:2*, Abstract DOI: 10.1080/03090728.2020.1809314.
   + Industrial Archaeology News, 188, Spring 2019, p. 8
   https://industrial-archaeology.org/wp-content/uploads/2020/10/AIA-News-188-Spring-2019.pdf
2. Industrial Archaeology News, 188, Spring 2019, p. 8
3. UNESCO Office Bangkok and Regional Bureau for Education in Asia and the Pacific (2020), p. 130
   https://www.google.co.uk/ books/edition/Asia_Conserved/0B0MEAAAQBAJ?hl=en&gbpv=1&dq=UNESCO+Wenzhou+alunite+mine+at+Fanshan&pg=PA131&printsec= frontcover
4. Souza, G. B. (2009) 'The French Connection: Indian Cottons, Their Early Modern Technology and diffusion', accessed via ResearchGate, p. 7
   https://www.researchgate.net/publication/268031253_The_French_Connection_Indian_Cottons_Their_Early_Modern_Technology_and_Diffusion/citation/download
   Gunn, G. (2011) *History without borders: The making of an Asian World Region, 1000-1800.* Hong Kong University Press, p. 144
   https://www.google.co.uk
5. Souza, G. B. (2009), p. 7
   Gunn, G. (2011), p 144
6. Guy, J. (2018) 'A Dutch Sarasa: The Global Journey of an Early Eighteenth-Century Coromandel Textile', *National Palace Museum Bulletin, vol. 50*, p. 4
   file:///C:/Users/User/Downloads/GUY_DutchSarasa_NPMB50_2018 .pdf

Riello, G. (2010) 'The Making of a Global Commodity: Indian Cottons and European trade, 1450-1850', *World History Studies and World History Education: The Proceedings of the First Congress of the Asian Association of World Historians* © 2010 The Asian Association of World Historians, p. 7
https://warwick.ac.uk/fac/arts/history/people/staff_index/griello/riello_giorgio.pdf

7. Arnolli, G. (2020) 'Chintz, new in the 17th century', International Council of Museums
https://costume.mini.icom.museum/wp-content/uploads/sites/10/2020/08/Gieneke-Arnolli-Chintz-new-in-the-17th-century.pdf
8. Arnolli, G. (2020) 'Chintz, new in the 17th century'
Lal, S. (2015) Guest Post: 'Indian Chintz – A Legacy of Luxury', *The V&A Blog*
https://www.vam.ac.uk/blog/fabric-of-india/guest-post-indian-chintz-a-legacy-of-luxury
9. Kotnola, S. (2017) 'Recreating the mystery of chintz', *The Asian Age, April 13* https://www.asianage.com/life/art/130417/recreating-the-mystery-of-chintz.html
10. Souza, G. B. (2009), p. 7
11. Riello, G and Roy, T. (2009) 'Introduction: The World of South Asian Textiles, 1500–1850', in:*How India clothed the World.* Leiden: Brill, p. 4
https://bhsecglobal.files.wordpress.com/2014/03/how-india-clothed-the-world-12.pdf
12. French Wikipedia: Histoire des indiennes de coton en Europe
13. Histoire du textile. episode 3, 24 May 2017: L'importation et la consommation des indiennes en France du XVII au XIXe, Radio France
https://www.franceculture.fr/emissions/la-fabrique-de-lhistoire/histoire-du-textile-3-limportation-et-la-consommation-des
14. Just History, April 14, 2018. https://upsc4history.wordpress.com/2018/04/14/calicos-a-medieval-craze/
15. The word *chintz* (derived from a Hindi word meaning spotted or speckled) originally denoted the very light printed cotton fabrics

so fashionable in the 18th century, but is now applied to any floral pattern, even printed on much heavier cloth.

16. Lal, S. (2015) Guest Post: 'Indian Chintz – A Legacy of Luxury', *The V&A Blog*
17. Sarah Fee, curator of an exhibition about chintz at the Royal Ontario Museum, quoted in
Bekhrad, J. (2020): 'The floral fabric that was banned', *BBC. The Collection*
https://www.bbc.com/culture/article/20200420-the-cutesy-fabric-that-was-banned
18. Riello, G. (2010), p. 14
19. Fisher, P. (2012) 'The Calico Acts: Why Britain Turned its Back on Cotton', *Sound Ideas, University of Puget Sound* p. 3 https://soundideas.pugetsound.edu/cgi/viewcontent.cgi?article=1001&context=history_theses
20. Sharpe, E. (2015) 'When the West went crazy for all things Asian', *the Art Newspaper*, 15 October 2015 https://www.theartnewspaper.com/2015/10/15/when-the-west-went-crazy-for-all-things-asian
21. Frances Wood, F., Roberts, G. (2005) 'Natural Fibers and Dyes', in *The Cultural History of Plants*. London: Routledge, p. 291
https://www.google.co.uk
22. Bekhrad, J. (2020): 'The floral fabric that was banned', *BBC. The Collection*
https://www.bbc.com/culture/article/20200420-the-cutesy-fabric-that-was-banned
23. Crosby, G. (2015) 'First Impressions: The Prohibition on Printed Calicoes in France, 1686-1759', thesis submitted at Nottingham Trent University for the degree of Doctor of Philosophy, p. 33. http://irep.ntu.ac.uk/id/eprint/28315/1/Gillian.Crosby-PhD%20thesis%202016.pdf
24. The trade lasted until the end of the 18th century when European mills learned to emulate Indian quality and introduced new techniques such as metal block printing in Switzerland.

Histoire du textile. episode 3, 24 May 2017: L'importation et la consommation des indiennes en France du XVII au XIXe, Radio France
https://www.franceculture.fr/emissions/la-fabrique-de-lhistoire/histoire-du-textile-3-limportation-et-la-consommation-des

Chapter 15

1. 'Caroline Years', pp. 61 and 63
   https://www.riksbank.se/globalassets/media/riksbanken-350-ar/tidslinjen/den-karolinska-tiden/46-74-caroline-years_eng.pdf
   Christina Piper (Tornflyt)
   http://historiska-personer.nu/min-s/p9d2c18f3.html
   Carlsson, K. 'Alum and Andrarum's Alunbruk
   https://alunbruket.com/alunbrukets-historia/
2. https://www.skaneleden.se/en/etapp/8-alunbruket-heinge
3. Winter 1740, grain harvests: Emmanuel Le Roy Ladurie, E. L-R, (2006) *Histoire humaine et comparée du climat Tome 2: Disettes et révolutions 1740-1860*. Paris: Fayard, pp. 1799, 1821 and 1828
   https://books.google.co.uk/
4. Baker Baker Papers, Scope and Content. Jisc Archives Hub.
   https://archiveshub.jisc.ac.uk/search/archives/7f9250b1-13bd-3c2f-9478-f52e503ea4e4
5. Beaumont, S. L. (2019), pp. 149 and 152
   Wiskin, C. (2003) 'Urban Businesswomen in 18th century England', Chapter 5 in *Women and Urban Life in 18th century England*, edited in Rosemary Sweet and Penelope Lane. London: Routledge,
   https://www.google.co.uk
6. Wiskin, C. (2003) 'Urban Businesswomen in 18th century England'
7. It ended in 1647 with the fall of Charles I.
8. Judith had her own legally protected separate estate, comprising £2,000 in land and property and a monthly allowance of £1,000.
   Beaumont, S. L. (2019), p. 149
9. Historic England Research Records: Boulby Alum Quarries and Works

https://www.heritagegateway.org.uk/Gateway/Results_Single.aspx?uid=612070&sort=4&search=all&criteria=steeping&rational=q&recordsperpage=60&resourceID=19191

10. Beaumont, S. L. (2019), pp. 147 and 153-155
11. Beaumont, S. L. (2019), pp. 153 and 157
    The Wikipedia article on Impressment gives a figure of 74, 771, taken from:
    Fischer, L. R. and Nordvik, H. W. (1990) *Shipping and Trade, 1750–1950: Essays in International Maritime Economic History*, p. 25.
    https://en.wikipedia.org/wiki/Impressment
12. ELMAP records database, Centre for Maritime Historical Studies, University of Exeter
    http://humanities.exeter.ac.uk/includes/centres/cmhs/ELMAP/location.php?loc_id=33
13. *Description of England and Wales, Vol. X,* (1770). London: Newbery and Carnan, p. 270
14. Sutherland, L. (1984) *Politics and Finance in the 18th century.* London: Hambledon Press, pp. 450 and 458
    https://www.google.co.uk
15. Beaumont, S. L. (2019), p.158
16. Beaumont, S. L. (2019), p.157
17. Cotterill, M. (2016), section 36
    The Northern Star or Yorkshire Magazine, January 1818, p. 266
    https://www.google.co.uk/books/edition/The_Northern_Star_Or_Yorkshire_Magazine/C4M4AQAAMAAJ?hl=en&gbpv=1&dq=Godeland+Bank+alum&pg=PA266&printsec=frontcover
18. The Northern Star January 1818, p. 266
19. Boydell, S. (2019) 'The Shah of Allum: Sir George Colebrooke', *Bath History Vol. 15*, p. 124
    https://historyofbath.org/images/BathHistory/Bath%20History%20Vol%2015%20-%20Stuart%20Boydell%20-%20The%20Shah%20of%20Allum-%20Sir%20George%20Colebrooke.pdf
20. 12000 quintals
21. The figure of 7000 quintals is given by Lucy Sutherland. 7000 quintals is equivalent to about 770 tons, and the estimated annual

production of alum at Karahissar in the eighteenth century was, according to Suraiya Faroqhi, approximately 750 tons.
Sutherland, L. (1984), p 453
Faroqhi, S., (1979), p. 164
Colebrook maybe negotiated with the merchants in Izmir rather directly with the alum producers because the camel handlers were notorious for stealing alum during transport to the coast.
Dingec, E., XVIII. VE XIX. Yüzyillarda Kütahya a Gediz (Gedos0 Şap Madeni, p. 150 https://dergipark.org.tr/en/download/article-file/1270516

22. Christian and Muslim powers in the Mediterranean, including Rome, continued to enslave captives till about 1825.
Wainwright, M. C. and Michelson, E. (2021) *A Companion to Religious Minorities in Early Modern Rome*. Leiden: Brill, pp. 299 and 310
file:///C:/Users/User/Downloads/ebin.pub_a-companion-to-religious-minorities-in-early-modern-rome-9789004443495.pdf
23. IL CASTELLO DI ROTA, Bolletino p 75
https://www.artestoriatarquinia.it/wp-content/uploads/bollettini/1992_Bollettino/Bollettino%201992.pdf
24. Sutherland, L. (1984), p. 453
25. Sutherland, L. (1984), pp. 456 and 460
26. Sutherland, L. (1984), p. 462
27. Colebrook had invested £123 000 in the scheme, with a further £43 000 owed to creditors in London, Venice, Danzig, and Bordeaux.
Sutherland, L. (1984), p. 461
28. Sutherland, L. S. (1936) 'Sir George Colebrooke's world corner in alum 1771-73', *Economic History, vol. 3, no. 11*, pp. 238-240. *JSTOR*, www.jstor.org/stable/45366462. Accessed 9 Aug. 2021.
Kosmetatos, P. (2014) Financial contagion and market intervention in the 1772-3 credit crash', *Working Paper No. 21, Darwin College, University of Cambridge*
http://www.econsoc.hist.cam.ac.uk/docs/CWPESHnumber21October2014.pdf
29. Boydell, S. (2019) The Shah of Allum: Sir George Colebrooke

30. Beaumont, S. L. (2019), p. 159
31. Appleton, P. in: Twigg, C. Alum Works in the Project Area, 14th February 2018
https://east-clevelands-industrial-heartland.co.uk/2018/02/14/alum-works-in-the-project-area/
32. Beaumont, S. L. (2019), pp.160-161, and 153
33. Northern Star of Yorkshire, April 1818 page 267
google.co.uk/books/edition/The_Northern_star_or_Yorkshire_magazine/XPcHAAAAQAAJ?hl=en&gbpv=1&dq= Eskdaleside%09%09270+Sandsend%09%09455+Kettleness%09%09455+Boulby%09%09%09450+Lofthouse%09%09910&pg=PA267&printsec=frontcover
34. Boydell, S. (2019), p. 127
35. Beaumont, S. L. (2019), p.147
36. Such was her expertise that Durham City Council sought her opinion her such things as the misuse of charity money and the appointment of wardens for city trades.
Beaumont, S. L. (2019), pp. 149 and 150

Chapter 16

1. These stones were, in fact, ammonites. In popular folklore they were held to be snakes turned into stone by St Hilda. Ammonites are still easily found on the Yorkshire coast.
Page, K.N. (2004) 'Whitby to Saltwick, North Yorkshire (NZ 901 115–NZ 916 109)', *JNCC British Lower Jurassic Stratigraphy, Chapter 6 – The Cleveland Basin*, Introduction, p. 299, http://www.habitas.org.uk/staff/staff_pdf_files/lowjurchap6.pdf
2. The Yorkshire fossil hunters were, as far as I know, all men. I have found no account of a Yorkshire Mary Anning (1799-1847), famous for her discoveries of fossils on the coast of Dorset.
http://www.bbc.co.uk/northyorkshire/content/articles/2005/07/21/coast05walks_stageseven.shtml
3. Osborne, R. (1999) *The Floating Egg.* London: Pimlico, pp. 24-25
4. Swallow Head Farm. Wider Historical Context, p.40
5. Osborne, R. (1999), p. 44

Benton, M. J, and Taylor, M. A. (1984) 'Marine reptiles from the Upper Lias (Lower Toarcian, Lower Jurassic) of the Yorkshire Coast', *Proceedings of the Yorkshire Geological Society, Vol. 44, Part 4, No 29*, pp. 399-429
http://pygs.lyellcollection.org/
Whitby's Teleosaurus Chapman, October 11, 2016, in *Whitby Uncovered*
https://whitbyuncovered.wordpress.com/2016/10/11/whitbys-teleosaurus-chapman/

6. Perrin, L. (2021) 'Rare ichthyosaur fossil found on beach at Kettleness, near Whitby', The Scarborough News
7. https://www.thescarboroughnews.co.uk/news/people/rare-ichthyosaur-fossil-found-on-beach-at-kettleness-near-whitby-3463165
Cotterill, M. (2016), section 36: Whitby Port and the alum industry
8. Bloom, J.M. (2017) *Papermaking: The Historical Diffusion of an Ancient Technique*, in: Jöns, H., Meusburger, P., Heffernan, M. (eds) Mobilities of Knowledge. Knowledge and Space, vol 10. Springer, Cham. https://doi.org/10.1007/978-3-319-44654-7_3
9. Whatmans and Wove Paper. http://www.wovepaper.co.uk/alumessay2.html. Appendix I, Part 4, footnote 10
10. Whatmans and Wove Paper
'J Whatman – the master of Western Papermaking', April, 2019. © 2022, Vintage Paper Co https://vintagepaper.co/blogs/news/drying-paper-at-whatmans-springfield-mill
11. 'Chiltern Straw Plaiters', *Aston Rowant & Chilterns Spring Line Villages Heritage & History*, https://astonrowant.wordpress.com/chiltern-hills/straw-plaiters/
12. White, W. (1861), pp. 97-103
Marx, K. (1867) *das Kapital*, Buch 1 [0157: 138]
Online Library of Liberty (libertyfund.org)
13. Bun, K. M. (2001) *The Salt Merchants of Tianjin*. Honolulu: University of Hawai'i Press, p. 48 https://scholarspace.manoa.hawaii.edu/bitstream/10125/23059/1/%2353_Bun.pdf

14. Tomkins, A. (2005) *Women and poverty in Women's history: Britain, 1700-1850.* Abingdon, Routledge, p. 158
    Hardy, A. (2003) 'Commentary: Bread and alum, syphilis and sunlight: rickets in the nineteenth century', *International Journal of Epidemiology, Volume 32, Issue 3*, pp. 337–340. https://doi.org/10.1093/ije/dyg175
15. Starmans, B. J. : Give Us This Day Our Daily Bread, *The Social Historian*
    https://www.thesocialhistorian.com/give-us-this-day-our-daily-bread/
    Hart, F. L. (1952) 'A history of the Adulteration of Food Before 1906', *Food, Drug, Cosmetic Law Journal*, *7*(1), 5–22. http://www.jstor.org/stable/26654178
    Gerhardt, K-U. (2020) 'Das Speenhamland-System als frühes Grundeinkommen. Vorgeschichte, Wirkungen und Erkenntnisse', Forschungsarbeit https://www.grin.com/document/950034
    Marx, K. (1867) *das Kapital*, Buch 1 [0157: 138]
16. Walton, G. (2014) 'Food and Drink Adulteration in the 1700 and 1800s', quoting Accum, F. C A. *A Treatise on Adulterations of Food, and Culinary Poisons,* 1820, p. 131.
    https://www.geriwalton.com/food-and-drink-adulteration-in-1700-and/
17. Tomkins, A. (2005) p. 158
    + Hardy, A. (2003) pp. 337–340
18. Russell, P. in Financial Times, 27 Sept. 2013 https://www.ft.com/content/2be889d8-257a-11e3-b349-00144feab7de
19. Anon (1830) *Deadly Adulteration and Slow Poisoning Unmasked.* London: Sherwood, Gilbert and Piper, p. 5
    https://www.gutenberg.org/cache/epub/52434/pg52434-images.html
20. '10 dangerous things in Victorian/Edwardian homes', BBC website, 16 December 2013
    https://www.bbc.co.uk/news/uk-25259505
21. Hardy, A. (2003) pp. 337–340

22. Chesney, R. W. (2012) 'Environmental Factors in Tiny Tim's Near-Fatal Illness', *Review, Archives of pediatrics & adolescent medicine,* pp. 271-275 https://jamanetwork.com/journals/jamapediatrics/fullarticle/1107722
23. History of Food Law: The 19th Century, 15 September 2014, *Artisan Food Law* http://www.artisanfoodlaw.co.uk/history-of-food-law/19th-century/history-of-food-law-19th-century
24. Winter, J. M. (1980) 'Military Fitness and Civilian Health in Britain during the First World War', *Journal of Contemporary History* 15, no. 2 pp. 211–44. http://www.jstor.org/stable/260511.
    + Walter W Holland, W. W. and Stewart, S. (1998) *Public health: The vision and the challenge.* London: The Nuffield Trust © Nuffield Trust pp. 24 and 25
    https://www.nuffieldtrust.org.uk/files/2017-01/public-health-vision-challenge-web-final.pdf
25. Wiesel, E., quoted in
    Skloot, R. (2011) *The Immortal Life of Henrietta Lacks.* London: Pan Books, 2011, Foreword.
26. Very useful local information about Skelton at Bill Danby's pages now run by the Skelton History Group at
    http://skeltonincleveland.com/
    Information about Eston, a village 8 miles away in blog by Kevin Dillow at
    https://landofiron.org.uk/collections-blog/eston-the-impact-of-ironstone-mining-in-the-first-decade-1850-1861
27. https://www.prisonhistory.org/prison/north-riding-house-of-correction-at-northallerton/
28. Loyd, C. (2020) 'The 'disgusting' hard labour treadmills at Northallerton Prison and how officers got the nickname *screws*', *The Northern Echo*
    https://www.thenorthernecho.co.uk/history/18252687.disgusting-hard-labour-treadmills-northallerton-prison-officers-got-nickname-screws/
29. Census returns:

http://wyndhammarsh.co.uk/genealogy/getperson.php?personID=I18548&tree=CledwynMarsh

30. https://www.geni.com/people/Elizabeth-Temple/6000000024288502764
    And https://www.geni.com/people/Jane-Temple/6000000029940693063
31. James in 1831, Joseph in 1833, John in 1836, Thomas in 1839, Adam in 1842 and William in 1845. James, Joseph, John and Adam became miners.
32. 'The Housing Act, 1957. Guisborough (Belmangate No. 2 Clearance Area) Compulsory Purchase Order 1960.'
    http://guisboroughhistorynotes.blogspot.com/2011/02/
33. Runswick: a tale of landslips – and the cholera of 1866: in *North Yorkshire History, from Hutton Rudby to Stokesley, Guisborough, Whitby ... and beyond the county*, Saturday, 12 December 2020 http://northyorkshirehistory.blogspot.com/

Chapter 17

1. *The Industry of Nations, as exemplified at the Great Exhibition of1851* (1852), London: Society for Promoting Christian Knowledge, accessed at: *SearchWorks Catalog, Stanford Libraries,* p. 344
   https://searchworks.stanford.edu/view/548309
2. White, W. (1861), p. 116
3. Stephenson, R. (1851) *The Great Exhibition*. London: Routledge, p. 35
   Picture of the alum pillar at http://gotomootpoint.blogspot.com/2015/08/lets-sneak-into-1851-worlds-fair.html
   Carroll, J. S, (2017) 'The Victorians taught children about consumerism – and we can learn from them too', *The Conversation,* https://theconversation.com/the-victorians-taught-children-about-consumerism-and-we-can-learn-from-them-too-76658
4. Cotterill, M. (2016), section 36
   Reports of the United States Commissioners to the Paris Universal Exhibition, 1878, Volume IV (Washington: Government Printing Office, 1880, p. 91, accessed at https://

www.google.co.uk/books/edition/Reports_of_the_United_States_Commissione/BvA_AAAAYAAJ?hl=en&gbpv=1&dq=%22Milos%22+%22alum%22&pg= PA91&printsec=frontcover

5. Cotterill, M. (2016), section 36: Scottish Alum Works
   Skillen, B. S. (1989) 'Aspects of the Alum Mining Industry about Glasgow', *British Mining No, 39, the Northern Mine Research Society, Sheffield*, pp. 53-60
   https://www.nmrs.org.uk/assets/pdf/BM39/BM39-53-60-alum.pdf
   Sinclair, R. 'Textile Dyeing in Paisley 1800-1840', *Renfrewshire Local History Forum Journal, © Renfrewshire Local History Forum & Authors*
   http://rlhf.info/wp-content/uploads/14.1-Textile-Dyeing-Sinclair.pdf
6. Le Grand, H. E. (1984). Theory and Application: The Early Chemical Work of J. A. C. Chaptal. *The British Journal for the History of Science, 17*(1), pp. 34 and 35. http://www.jstor.org/stable/4026015
7. Souza, G. B. (2009), pp. 11 & 12.
   Spence later opened a second works at Goole in Yorkshire.
   Rickard, D. (2015), p. 41
   Cotterill, M. (2016), section 36: Spence's Pendleton Alum Works in Lancashire
8. Appleton, P. (2018) 'Alum Works in the Project Area', *article in East Cleveland's Industrial Heartland*
   https://east-clevelands-industrial-heartland.co.uk/2018/02/14/alum-works-in-the-project-area/
   'Alum quarrying: The effect of extractive industry on the Yorkshire Coastal Landscape' https://web.archive.org/web/20110716030245/http://www.fortunecity.com/greenfield/ecolodge/25/alum.htm
   Young 1817, quoted in: Swallow Head Farm. Wider Historical Context, p.26
   Schofield, P. (2007) 'National Trust Holdings on the North Yorkshire Coast', *Historic Landscape Survey* Section 4.4.7 https://library.thehumanjourney.net/4542/1/Full_Yorks_Coast_Rep.pdf
   Twigg, C. (2009) 'Alum House and Cementstone Kiln – Hummersea Steps', *post on Hidden Teesside*

http://www.hidden-teesside.co.uk/2009/09/23/alum-house-hummersea-steps/
Boulby Alum Quarries and Works, *Historic England Research Records*
https://www.heritagegateway.org.uk/Gateway/Results_Single.aspx?uid=612070&sort=4&search=all&criteria=steeping&rational=q&recordsperpage=60&resourceID=19191

9. The alum was made here from the ash created by burning the local lignite (brown coal).
'Projekt *Kohle und Alaun*', Denkmal und Geschichtsverein, Bonn Rechtsrheinisch e. V.
https://www.denkmalverein-bonn.de/projekte-kohle-alaun
10. Morassina alum mine at Schmiedefeld, Jeremias Glück mine at Garnsdorf, near Saalfeld
11. Christian Gottlob Frege, *Leipziginfo*
https://www.leipziginfo.de/beruehmte-leipziger-personen/person/frege/
'Revier Saalfeld', *Bergmannsverein Erfurt e.V.*
https://www.bergmannsverein-erfurt.de/main.php?unmenue=bergbau_frueher_thueringen&content=saalfeld
12. Publicity website.
https://www.vogtland-tourismus.de/de/poi/bergwerk/alaunbergwerk-ewiges-leben-plauen/7014436/
13. Wohlfarth, B. (2020) 'They come and go', *Tag Archives: Alum Shale*
https://barbarawohlfarth.wordpress.com/tag/alum-shale/
14. Date de soumission: 03/09/2015
https://whc.unesco.org/fr/listesindicatives/6067/
15. Stasolla, F. R. (2020) 'The allumiere of the Tolfa district: archaeology and economical network'
Aygün, C. Ö. (2020), p. 175
16. Michael Hollington, M. and Orestano, F. (2009), Introduction p. xviii
17. The former alum works lie partly under Allumiere's *campo sportive*, its sports ground.
Re forest:
Jones, A. (2016)

18. Jeremias Glück and Morassina cater for people with breathing difficulties, who can benefit from the germ- and allergen-free air inside the mine.
    Publicity website: https://www.thueringen.info/morassina-grotte.html
19. Bucan, S., (2019), 'Badespaß im *Toten Meer* von Böhmen', Radio Prague International, Sendung auf Deutsch
    https://deutsch.radio.cz/badespass-im-toten-meer-von-boehmen-8124400
20. Uranium, oil and vanadium have been extracted from the alum shale waste at Kinnekulle, Kvarntop and Andrarum.
    Wohlfarth, B. (2019) 'History is talking', *Thoughts, Tales and Travels*
    https://barbarawohlfarth.wordpress.com/2019/06/28/history-is-talking/
21. Britton, N. (2014) 'Alunite', *Ghost Towns USA Canada*
    https://www.ghosttowns.com/states/ut/alunite.html
    Kennedy, R. R. (1963) *Geology of Piute County, Utah* in Marysvale Alunite, April 2020.
    https://utahrails.net/mining/marysvale-alunite.php
22. Feng, S., Qian, W., Sanchiz, J. M. C., White, R & Robinson, M. (2020), pp. 98-113
    'Fudewan Miners Village in Wenzhou Wins UNESCO Award' (2016)
    http://english.wenzhou.gov.cn/art/2016/9/5/art_1231337_3999844.html

Chapter 18

1. 'Alexander, Freiherr von Humboldt 1792 auf Inspektion im Schmiedefelder Alaun- und Vitriolwerk', *Außerschulische Lernorte in Thüringen*
   file:///C:/Users/sydth/Downloads/Lernort_Morassina_Humboldt_Informationsblatt_Humboldt_Schaubergwerk_Morassina%20(1).pdf
2. 'Humboldtscher Licht-Erhalter', *Sammlungen der TU Bergakademie Freiberg Sammlung für Bergbaukunde [IX. D 1]*
   https://nat.museum-digital.de/object/1089778?navlang=de
   https://www.frankentourismus.de/poi/bergwerk_bestaendiges_glueck-33150/

3. Humboldt, A. and Bonplant, A. (1826) *Personal Narrative of travels to the equinoctial regions of the new continent, during the years 1799-1804*. London: Longman, Rees, Orme, Brown, and Green, pp. 92 to 104
http://darwin-online.org.uk/content/frameset?pageseq=27&itemID=A597.6a&viewtype=side
4. Antúnez de Mayolo, K. K. (1989) 'Peruvian Natural Dye Plants', *Economic Botany, vol. 43, no. 2*, p. 182
http://www.jstor.org/stable/4255151. Accessed 12 Apr. 2022.
5. Colouring cloth: How the Vikings dyed their textiles, *Follow the Vikings, Cultural Route of the Council of Europe*
https://www.followthevikings.com/discover/culture/textiles/how-vikings-coloured-cloth
Jenny Dean's Wild Colour, two blogs (2009), accessed at:
https://www.jennydean.co.uk/anglo-saxon-dye-experiments-part-1/
https://www.jennydean.co.uk/anglo-saxon-mordants/
Stasińska, K. 'Before They Dyed. Mordants and Assists in the Textile Dyeing Process in Anglo-Saxon and Anglo-Scandinavian Britain: An Experimental Approach;, reviewed Article: 2021-11-25, *EXARC Journal Issue 2021/4*. https://exarc.net/ark:/88735/10603
Vajanto, K. and Pasanen, M. 'Lichen Purple, Tannin Red Reconstructing a Viking Age Finnish woman's woollen shawl', Department of Applied Physics, Aalto University, Finland
https://research.aalto.fi/en/publications/lichen-purple-tannin-red-reconstructing-a-viking-age-finnish-woma/fingerprints/
America:
https://www.suzannedekel.com/post/using-clubmoss-as-an-alum-mordant
6. The Outside, Natural Authentic Colours (2017), *Monthly Archives*
https://www.theoutside.org.uk/2017/01/
7. Marco Schmitt, M., Boras, S., Tjoa, A., Watanabe, T., and Jansen, S, (2016) 'Aluminium Accumulation and Intra-Tree Distribution Patterns in Three Arbor aluminosa (Symplocos) Species from Central Sulawesi', *Peer-reviewed research Article, Plos One Journal*
https://doi.org/10.1371/journal.pone.0149078

Index